NEW DIRECTIONS IN ATOMIC PHYSICS

VOLUME II: EXPERIMENT

YALE SERIES IN THE SCIENCES

NEW DIRECTIONS IN ATOMIC PHYSICS

VOLUME II: EXPERIMENT

EDITORS: EDWARD U. CONDON AND OKTAY SINANOĞLU

NEW HAVEN AND LONDON, YALE UNIVERSITY PRESS, 1972

Library of Congress catalog card number: 78-140542.
International standard book number: 0-300-01400-7.

Designed by Sally Sullivan
and set in Monophoto Times Roman type.
Printed in the United States of America by
The Murray Printing Co., Forge Village, Mass.

Distributed in Great Britain, Europe, and Africa by
Yale University Press, Ltd., London;
in Canada by McGill-Queen's University Press, Montreal;
in Latin America by Kaiman & Polon, Inc., New York City;
in Australasia by Australia and New Zealand Book Co.,
Pty., Ltd., Artarmon, New South Wales;
in India by UBS Publishers' Distributors Pvt., Ltd., Delhi;
in Japan by John Weatherhill, Inc., Tokyo.

Contents

Contents of Volume I: Theory

Preface

Atomic physics has seen a strong revival in the last decade. Major developments have taken place and are taking place in both theory and experiment. Some of these are covered in these two volumes by our esteemed colleagues who have been prime contributors to the special areas involved. It is hoped that succeeding volumes may be published in the Yale Series covering other major areas, which of course could not be covered in two brief volumes.

The topics in the present two volumes were chosen initially to be presented as lectures to about ninety researchers and graduate students from sixteen countries. A NATO Advanced Study Institute was organized in İzmir, Türkiye. Discussions among the lecturers and other participants contributed to stimulate interest in combining in the future the different approaches in some of the theoretical areas to make the results applicable to a wider range of problems. Such potentialities appear to exist, for example, across group theory, N-electron correlation theory, theory of relativistic effects, and methods for treating electron-atom scattering phenomena. The various results in two volumes are, of course, of practical use, because of the nature of atomic physics, in a wide range of fields like astrophysics, theoretical chemistry, and atmospheric physics.

Our thanks are due to the lecturers for their excellent work in İzmir and in their writing, to the NATO Science Bureau, the Turkish Scientific and Technical Research Council (Türkiye Bilimsel ve Teknik Araştirma Kurumu) and the Orta-Doğu Teknik Üniversitesi (ODTÜ) of Ankara, Türkiye, for financial support of the Institute in İzmir, and to Prof. I. Rabi, Prof. L. T. Muus, Prof. V. Hughes, and Dr. V. W. Cohen for their interest in the Institute. We thank also Prof. Yusuf Vardar of Ege Üniversitesi, İzmir, Prof. Erdal İnönü of ODTÜ, Dr. İskender Öksüz, and Dr. Halis Odabaşı, without whose help and interest the lectures in İzmir could not have been arranged. Mr. Hikmet Özdizlây of Ege and Mr. Tâcan Önder of ODTÜ were also very helpful during the Institute. Finally our thanks go to Mr. Chester Kerr, Mr. Dana Pratt, and Mrs. Anne Wilde and our many other friends at the Yale Press for their interest in these books.

Boulder and New Haven   Edward U. Condon
1970   Oktay Sinanoğlu

Contributors

R. H. GARSTANG
Joint Institute for Laboratory Astrophysics of the
National Bureau of Standards and the University of Colorado
Department of Physics and Astrophysics
University of Colorado
Boulder, Colorado, U.S.A.

ALFRED KASTLER
Laboratoire de Physique
Ecole Normale Supérieure
Université de Paris
Paris, France

RICHARD MARRUS
Department of Physics and Lawrence Radiation Laboratory
University of California
Berkeley, California, U.S.A.

CLEANTHIS NICOLAIDES
Sterling Chemistry Laboratory
Yale University
New Haven, Connecticut, U.S.A.

ADNAN SAPLAKOĞLU
Department of Physics
Orta-Doğu Teknik Üniversitesi
Ankara, Türkiye

OKTAY SINANOĞLU
Sterling Chemistry Laboratory
Yale University
New Haven, Connecticut, U.S.A.

1

ALFRED KASTLER

The Principles of Optical Pumping

PART 1. NONTHERMAL DISTRIBUTIONS IN ATOMIC STATES

Optical pumping is a method to produce, by light irradiation, nonthermal distributions of populations among atomic energy states.

NONTHERMAL DISTRIBUTION IN EXCITED STATES OF ATOMS

If atoms are excited in an ordinary gas discharge by isotropic electron impact, the Zeeman levels of the excited states are equally populated and the spectral lines emitted from these levels are nonpolarized. But if the exciting conditions are anisotropic or dissymmetric, the Zeeman levels of the excited states are unequally populated, and the light emitted is more or less polarized.

Anisotropic exciting conditions are obtained by optical excitation by a light beam directed in space, either nonpolarized or polarized linearly or circularly. Excitation is also anisotropic by electron impact, if it is caused by a beam of electrons directed in space. We may now study examples.

The Brossel–Bitter Double Resonance Experiment on State 6^3P_1 of the Mercury Atom

The ground state of the mercury atom is a state 6^1S_0 of the electron configuration. It has spherical symmetry, and its magnetic moment is zero. It is a diamagnetic state. By optical excitation, by irradiation of mercury vapor with the ultraviolet line 2537 Å emitted by a mercury light source, the atoms are brought to excited state 6^3P_1 (Fig. 1-1). If the vapor pressure is very small so that collisions in the excited state may be neglected, after a mean life time of 10^{-7} sec, the atoms fall to the ground state, emitting photons of 2537 Å in all space directions. This is Wood's famous experiment of optical resonance (Fig. 1-2). The light emitted is polarized. This polarization is explained by the Zeeman scheme of line 2537 Å. The upper state 6^3P_1 with $J = 1$ is paramagnetic and is

1

Alfred Kastler

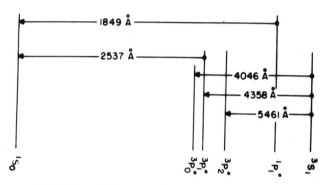

Fig. 1-1. Part of the energy-level diagram of the Hg atom.

split in a magnetic field H into three magnetic sublevels $m_J = -1, 0$, and $+1$ with two equal intervals proportional to field H_0:

$$\Delta E = \hbar(\omega_0) = \hbar(\gamma)H_0$$

The term γ is called the gyromagnetic ratio of the state and is related to the Landé g factor by

$$\gamma = g\frac{e}{2m_0}$$

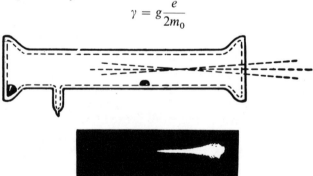

Fig. 1-2. R. W. Wood's experiment of optical resonance of mercury vapor. Photograph of the scattered resonance light 2537 Å.

where e is the charge and m_0 the mass of the electron. We can introduce the Bohr magneton $\mu_B = \hbar(e/2m_0)$ and write:

$$\Delta E = g(\mu_B)H_0$$

$g(\mu_B)$ being the magnetic moment associated with state 6^3P_1. Line 2537 Å is split into three Zeeman components, as shown by Figure 1-3.

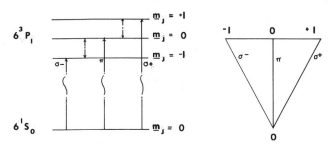

Fig. 1-3. Energy scheme (left) and polarization scheme (right) of the Zeeman structure of mercury-2537.

The three components differ in frequency. (The line is split into a triplet in a strong magnetic field where the Zeeman interval is larger than the Doppler width of the line.) The components differ also in polarization: σ^-, π, σ^+.

If the incident-exciting light is polarized π (linearly) or σ^+ or σ^- (circularly), selective excitation of only one Zeeman level of state 6^3P_1 is obtained, and the reemitted resonance light has the same polarization as the incident light (depending on the direction of observation).

This conclusion remains true for a small H_0 field where the Zeeman components are not separated and even for a zero field. (The magnetic earth field has to be exactly compensated.)

To Wood's device, Brossel (at Massachusetts Institute of Technology with Bitter, 1949–1951) added a radiofrequency (RF) coil producing alternative magnetic field H_1 (cos ωt) perpendicular to H_0.

If the resonance condition $\omega = \omega_0 = \gamma H_0$ is fulfilled, magnetic dipole transitions are induced between Zeeman levels of the excited state. This is called magnetic resonance. These transitions equalize the populations of all m_J states. The emitted resonance line is depolarized, for example, if the excitation is π excitation. Without magnetic resonance, the emitted light is π polarized. By applying magnetic resonance, we find that the σ components appear in the emitted light. Magnetic resonance is monitored by measuring $\mathcal{I}_\sigma$, the intensity of emitted σ light.

Figure 1-4 shows magnetic resonance curves obtained by Brossel; ω is kept constant, equal to 144 MHz. The sweep through resonance is done by slowly changing H_0. The center of resonance is located at $H_0 = 69$ gauss. This corresponds to $g = 1.48$.

The theoretical value for Russell–Saunders coupling is $\frac{3}{2}$. Each resonance curve corresponds to a constant value of amplitude H_1 of the RF field (Majorana–Brossel curves). If H_1 grows, the resonance becomes larger. Extrapolation to $H_1 = 0$ gives the "natural resonance width." This extrapolation is made by plotting the square of the half-width $(\Delta H_0)^2$ as a function of H_1^2. From $\Delta\omega_0 = \gamma\Delta H_0$ we obtain $\Delta\omega_0 = 2/\tau$ according to the Heisenberg uncertainty relation (Fig. 1-5).

From this extrapolation the life time τ of the excited state 6^3P_1 is obtained. The measurement gives $\tau = 1.18 \times 10^{-7}$ sec.

From double-resonance experiments (optical resonance + magnetic resonance) g factors and life times of excited states can be obtained. Brossel himself observed magnetic resonance on 6^3P_1 of the odd mercury isotopes ^{199}Hg $(i = \frac{1}{2})$ and ^{201}Hg $(i = \frac{3}{2})$ whose g factors are quite different from the even ones, depending on the hyperfine coupling $\mathbf{F} = \mathbf{J} + \mathbf{i}$ of the states. The g factors measured in small fields are the g_F factors.

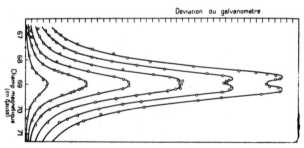

Fig. 1-4. Magnetic resonance curves of the excited state 6^3P_1 of Hg (Brossel 1952; Brossel and Bitter 1952).

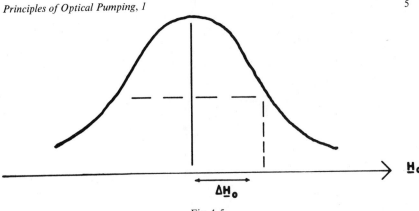

Fig. 1-5.

In case of nonzero nuclear spin and of hyperfine structure by optical excitation with polarized resonance light, nonthermal occupations of the F levels of the excited states are obtained. The hyperfine resonances $F_1 \rightarrow F_2$ of excited states can be monitored by the change of intensity and the polarization of the emitted light. In this way, hyperfine intervals between F states have been measured, especially for all alkali atoms.

These measurements have been made especially by Dr. Series in Oxford and by Professor Kopfermann's research group in Heidelberg. As a result the Q values of the electric quadrupole moments of the corresponding nuclei have been determined.

Figure 1-6 shows the mounting of a double-resonance experiment for Rb and Figure 1-7 gives the scheme of hyperfine structure for ^{87}Rb and ^{85}Rb. Figure 1-8 shows three hyperfine resonance maxima obtained for

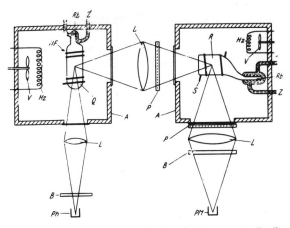

Fig. 1-6. Double resonance apparatus for the alkali vapors (zu Putlitz 1965).

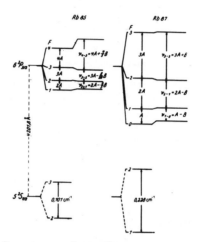

Fig. 1-7. Hyperfine structure of state $6^2P_{\frac{3}{2}}$ and ground state $5^2S_{\frac{1}{2}}$ of Rb.

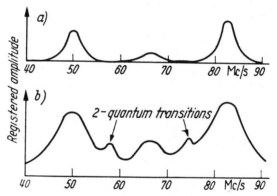

Fig. 1-8. Radiofrequency of resonances of hyperfine intervals of ^{133}Cs, state $7^2P_{\frac{3}{2}}$ (Althoff 1955).

^{133}Cs, and Figure 1-9 shows the corresponding hyperfine structure (hfs) involving hyperfine levels F = 2, 3, 4, and 5.

In this way the resonances F = 4 → F = 5 for the radioactive isotopes ^{135}Cs and ^{137}Cs have been determined by the Heidelberg group (Fig. 1-10).

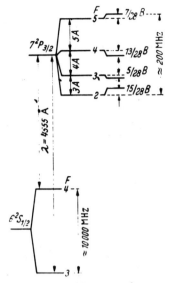

Fig. 1-9. Hyperfine structure of ^{133}Cs state $7^2P_{\frac{3}{2}}$ and ground state $6^2S_{\frac{1}{2}}$.

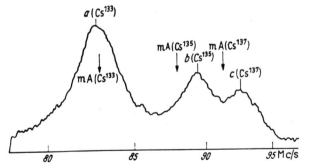

Fig. 1-10. Hyperfine resonances of $7^2P_{\frac{3}{2}}$, F = 5 → F = 4 for three Cs isotopes: 133, 135, 137 (Bucka, Kopfermann, and Otten 1959).

Table 1-1 gives a summary of the results obtained for all alkali nuclei (zu Putlitz 1965).

Table 1-1. Compilation of hfs measurements for alkali isotopes (zu Putlitz 1965)

Nucleus	State	Meth.	$a_{Mc/sec}$	$b_{Mc/sec}$	$Q_{10^{-24} cm^3}{}^c$	Ref.[d]
^{23}Na	$3\,^2P_{\frac{3}{2}}$	a*	19.5(6)	2.4(1.4)	+0.10(6)	56
		b	18.5(6)	2.25(40)	+0.097(13)	49
^{39}K	$4\,^2P_{\frac{3}{2}}$	a	6.20(12)	1.0(3)	+0.13(4)	57
	$5\,^2P_{\frac{3}{2}}$	a, b	1.97(10)	1.7(3)	+0.11(2)	48, 58
	$5\,^2P_{\frac{1}{2}}$	b	8.99(15)			50
^{40}K	$5\,^2P_{\frac{3}{2}}$	a	−2.450(46)	−1.31(33)	−0.093(25)	59
^{85}Rb	$5\,^2P_{\frac{3}{2}}$	a*	25.029(16)	26.032(70)	+0.298(1)	31, 51
	$6\,^2P_{\frac{3}{2}}$	a	8.16(6)	8.40(40)	+0.29(2)	60, 61
		a	8.178(9)	8.199(40)	+0.286(1)	52
		b	8.25(10)	8.16(20)	+0.283(8)	31
^{87}Rb	$7\,^2P_{\frac{3}{2}}$	a	3.72(3)	3.65(10)	+0.280(6)	53
	$5\,^2P_{\frac{3}{2}}$	a*	84.852(30)	12.611(70)	+0.144(1)	31
	$6\,^2P_{\frac{3}{2}}$	a	27.63(10)	4.06(20)	+0.14(1)	60, 61
		a	27.707(15)	4.000(39)	+0.140(1)	52
		a	27.70(2)	3.94(4)	+0.138(1)	55
^{133}Cs	$7\,^2P_{\frac{3}{2}}$	a	12.58(2)	1.72(4)	+0.133(1)	53
	$7\,^2P_{\frac{1}{2}}$	a	16.60(1)	−0.11(8)	−0.003(2)	42, 62
		a	16.609(5)	−0.16(6)	−0.0036(13)	45, 63
	$7\,^2P_{\frac{1}{2}}$	a	100.2(3)			46
	$8\,^2P_{\frac{3}{2}}$	a	7.626(5)	−0.049(42)	−0.0024(20)	64
^{135}Cs	$7\,^2P_{\frac{3}{2}}$	a	17.576(6)	2.19(9)	+0.049(2)	45, 63
^{137}Cs	$7\,^2P_{\frac{3}{2}}$	a	18.280(6)	2.23(9)	+0.050(2)	45, 63

[a] zero magnetic field.

[b] intermediate or strong magnetic field.

[c] Without Sternheimer correction.

[d] These references are found in zu Putlitz' article.

* These states were also investigated by atomic-beam resonance. See references 75–77.

Nonthermal Distribution in Excited States Obtained by Electron Impact

In small magnetic fields the Zeeman intervals between m states are small compared to the thermal energy kT. A thermal distribution corresponds in this case to equal populations of all m states.

A nonthermal distribution is obtained in two different ways. First, positive and negative m states have different populations. If this is the case, we speak of "atomic orientation." The simplest case is a $J = \frac{1}{2}$ state where there are only two m states: $m = -\frac{1}{2}$ and $m = +\frac{1}{2}$.

In one state the magnetic moment μ of the atom is parallel to the magnetic field H_0. In the other state it is antiparallel. Unequal populations of the two states result in the production of a bulk magnetic moment $\mathcal{M}$ of the vapor:

$$\mathcal{M} = \mu(N_+ - N_-)$$

N_+ being the population of state $m = +\frac{1}{2}$ and N_- of state $m = -\frac{1}{2}$. In the general case we define a *degree of orientation P* by:

$$P = \frac{\Sigma N_m \cdot m}{J \Sigma N_m}$$

Orientation can be obtained by irradiation with *circularly* polarized light. By absorption of this light the *angular moment* of the photons is transferred to the atoms.

Second, there is no difference of population between positive and negative m states of the same $|m|$, but different m states have different population. In this case we speak of *atomic alignment*.

In Brossel's experiment only state $m = 0$ of 6^3P_1 is populated. States $m = \pm 1$ remain empty. This is a case of alignment. It is obtained by excitation with *linearly* polarized light, and it gives rise to emission of *linearly* polarized light.

In the case $J = \frac{3}{2}$ we have alignment if the populations of states $m = \pm\frac{3}{2}$ are different from those of states $m = \pm\frac{1}{2}$. In the general case the *degree of alignment A* is defined by:

$$A = \frac{3}{J(2J - 1)}\left[\frac{\Sigma m^2 N_m}{\Sigma N_m} - \frac{J(J + 1)}{3}\right]$$

Alignment is obtained by optical space-directed excitation with linearly polarized light or with nonpolarized light or by space-directed electron-impact excitation. The light emitted shows linear polarization. In these cases magnetic resonance is monitored by the depolarization effect of the emitted spectral lines. In the case of electron-impact excitation where several lines are emitted, each line can be isolated by a monochromator and the resonance in its upper state can be observed. We must be aware of cascade effects (transfer of alignment).

Examples

Figure 1-11 shows the polarization of spectral lines of ^{4}He, observed by electron-impact excitation.

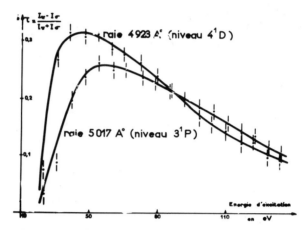

Fig. 1-11. Polarization of ^{4}He lines excited by electron impact.

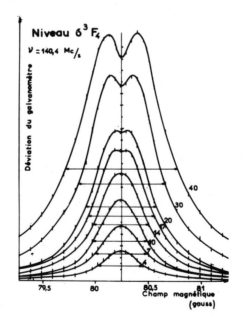

Fig. 1-12. Magnetic resonance curves of Hg atoms excited by electron impact (Pébay-Peyroula 1959).

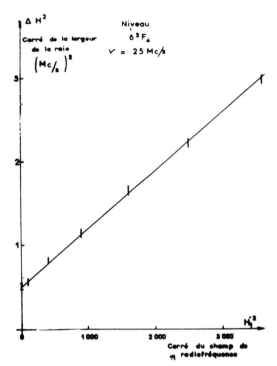

Fig. 1-13. Life time measurement of level 6^3F_4 of Hg; $v = 25$ Mc per sec (Pébay-Peyroula 1959).

Figure 1-12 shows magnetic resonance curves obtained on state 6^3F_4 of mercury excited by electron impact. These curves are analogous to Brossel's curves. The extrapolation of the resonance width leads to determination of the life time τ (Fig. 1-13).

Figure 1-14 shows some of the energy levels of the helium atom. Exciting He gas at very small pressure by an electron beam, Pébay–Peyroula was able to trace the magnetic resonance curve of state 3^3P of ^4He, to measure its g factor and life time (Figs. 1-15 and 1-16).

PRODUCTION OF A NONTHERMAL DISTRIBUTION IN GROUND STATES OF ATOMS

By irradiating atoms of an atomic beam or of a low-density vapor with circularly polarized light, orientation effects in the ground state are obtained. We will illustrate it on the odd mercury isotope ^{199}Hg which

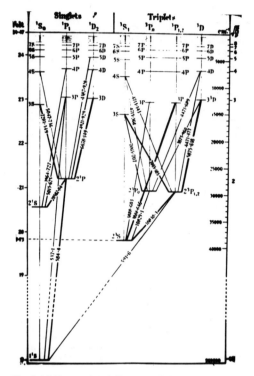

Fig. 1-14. Energy-level diagram of the helium atom.

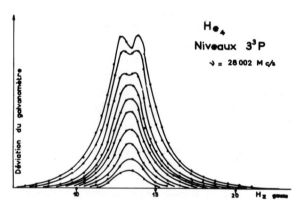

Fig. 1-15. Magnetic resonance curves of ^{4}He excited by electron impact, levels $3\,^3P_1$ and $3\,^3P_2$; $\nu = 28.002$ Mc per sec (Descoubes 1964, 1965).

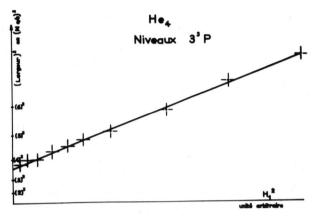

Fig. 1-16. Measurement of life time of level 3 ^{3}P of ^{4}He (Descoubes 1964, 1965).

has nuclear spin $i = \frac{1}{2}$. In the ground state 6^1S_0 with no electronic moment, the small nuclear magnetic moment associated with the spin leads, in a magnetic field H_0, to two nuclear magnetic substates $m_i = -\frac{1}{2}$ and $m_i = +\frac{1}{2}$ with opposite direction of the i vector. In the excited state 6^3P_1 the electronic angular moment **J** (in units $\hbar$) is coupled to the nuclear spin **i** by the hyperfine coupling a**IJ**. From this coupling two hyperfine states $F_1 = J - i = \frac{1}{2}$ and $F_2 = J + i = \frac{3}{2}$ result, separated by an interval of 0.7 cm^{-1} or 22.000 MHz.

Figure 1-17 in its upper left shows the energy level diagram of ^{199}Hg: the line 2537 Å is split into two hyperfine components, A leading to

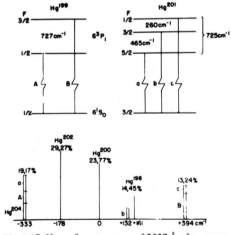

Fig. 1-17. Hyperfine structure of 2537 Å of mercury.

level $F_1 = \frac{1}{2}$ and B leading to level $F = \frac{3}{2}$ from the ground state. In the lower part of the figure the hyperfine structure of 2537 of natural mercury is shown. Each even isotope produces one single line, and the lines are displaced by isotope shift with a nearly constant interval of about $0.17\ \mathrm{cm}^{-1}$: 204, 202, 200, 198. Component A of ^{199}Hg is coincident with the line of ^{204}Hg; component B is far on the right in the frequency scale. This coincidence allows a selective excitation of level $F = \frac{1}{2}$ of ^{199}Hg by putting ^{204}Hg in the exciting light source and ^{199}Hg in the resonance cell. If the incident light is circularly polarized, a further selection occurs: Figure 1-18 shows the Zeeman scheme of component A of ^{199}Hg, showing four Zeeman components, two of polarization π, one σ^+, one σ^-. If the exciting light is σ^+ polarized, it can be absorbed only by atoms in the nuclear sublevel $m_i = -\frac{1}{2}$ of the ground state. The absorption leads to $m_F = +\frac{1}{2}$ of the excited state. (The positive angular momentum of the σ^+ photon is transferred to the atom.) The excited atom can fall back to the ground state in two different ways: return to state $m_i = -\frac{1}{2}$ by emission of a σ^+ photon or transit to state $m_i = +\frac{1}{2}$ by emission of a π photon. The ratio of transition probabilities of these two ways is 2 to 1. Each time three photons are absorbed, one of the three excited atoms is pumped from the $m_i = -\frac{1}{2}$ state to the $m_i = +\frac{1}{2}$ state. If this process of optical pumping is repeated a large number of times, the number N_- of atoms in state $m_i = -\frac{1}{2}$ decreases, the number N_+ of atoms in state $m_i = +\frac{1}{2}$ increases. This pumping process is counteracted by a relaxation process (when the Hg atoms collide with the silica walls).

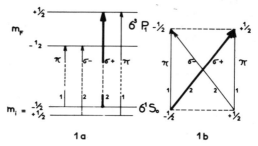

Fig. 1-18. Optical pumping schemes with circularly polarized light of ^{199}Hg, line 2537 Å; hfs component A ($i = \frac{1}{2} \rightarrow F = \frac{1}{2}$). (a) Energy scheme; (b) polarization scheme.

How can the progress of optical pumping be seen? Very simply. The incoming σ^+ photons can be absorbed only by atoms in state $-\frac{1}{2}$. If N_- decreases, the absorption of light decreases. During the optical pumping process the vapor becomes more and more transparent. The accompanying figures show the evolution in function of time. Figure 1-19 is the photograph of a trace of an oscilloscope spot swept from left to right as a linear function of time. The ordinate is the intensity of the pumping light transmitted through the vapor. At time $t = 0$ a shutter is removed.

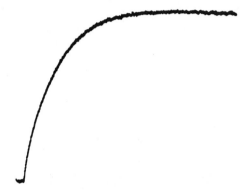

Fig. 1-19. Optical pumping: transient signal of transmitted light.

The light comes in, and its intensity is growing and reaching exponentially a steady-state value. In this steady state we have dynamic equilibrium between pumping and relaxation.

Figure 1-20 shows the transient signal of the intensity of the optical resonance light scattered at right angle of the incident beam. The second

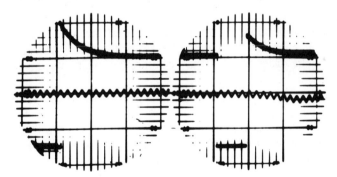

Fig. 1-20. Optical pumping of ^{199}Hg: transient signal of scattered resonance light (Cagnac 1960).

spot gives the time scale. As reemission is proportional to absorption, the intensity of this light decreases as a function of time. In the right-hand part of the figure the shutter has been inserted again for a small time interval after which it is again removed. During this interval of darkness the oriented nuclei have partly relaxed (returned from state $+\frac{1}{2}$ to state

$-\frac{1}{2}$). This experiment made by Cagnac is repeated several times, the darkness interval Δt being changed. The curves of the spot in these succeeding experiments are superposed on the same photograph as that shown in Figure 1-21. The starting points of the pumping curves lie on an exponential: e^{-t/T_1}. This leads to the determination of the constant T_1 which is the longitudinal nuclear relaxation time. (The magnetic moment which relaxes is "longitudinal," parallel to the applied magnetic field H_0.)

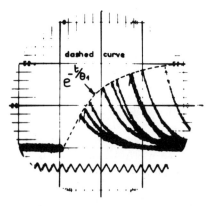

Fig. 1-21. Measurement of nuclear relaxation time T_1 in ^{199}Hg (Cagnac 1960).

At room temperature, T_1 turns out to be of the order of 3 to 5 sec. It is long. The length of the time of flight of the atoms in a cell of 2 to 3 cm^3 is of the order of 10^{-4} sec. Thousands of wall collisions are then needed before the nuclei relax.

Cagnac has measured T_1 also for ^{201}Hg, where it is about 100 times shorter.

Why is the relaxation much stronger for ^{201}Hg than for ^{199}Hg? Now ^{201}Hg has nuclear spin $i = \frac{3}{2}$. With this spin is associated a magnetic moment (smaller than the moment of ^{199}Hg) and a nuclear electric quadrupole moment. In the case of isotope 201, the dominant relaxation process is the interaction of this quadrupole moment with electric fields near the wall. This type of relaxation is absent for isotope 199 (nuclear spin = $\frac{1}{2}$), which has no quadrupole moment. Relaxation is strongly influenced by the mercury drop in the tail of the cell. To make this influence negligeable, a capillary tube is inserted between the cell and the tail. By sealing the tail, "dry mercury vapor" is produced and can be conserved for a long time in a fused silica cell.

Cagnac has studied the influence of the wall temperature on the relaxa-tion time T_1 of dry mercury vapor.

Figure 1-22 shows how T_1 depends on the temperature in the interval 20°C to 300°C for [199]Hg. The higher the temperature the smaller the relaxation. The results suggest that the mercury atoms stick to the wall of fused silica and that the mean sticking time depends strongly on the temperature. A quantitative study enabled Cagnac to measure the sticking energy. It is surprisingly high: 0.18 electron volt.

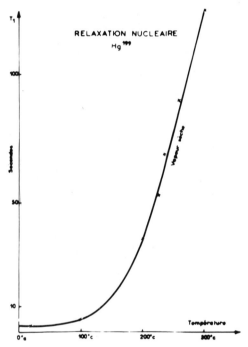

Fig. 1-22. Relaxation time T_1 versus temperature for [199]Hg (Cagnac 1960).

Recently Cagnac and Lemeignan extended the measurements to 600°C for both isotopes 199 and 201. Figures 1-23 and 1-24 show the results, Figure 1-23 with a fresh cell, Figure 1-24 after many reversible temperature cycles have been made.

Note that at high temperature the relaxation rate for [201]Hg becomes smaller than for [199]Hg. (The quadrupole relaxation process becomes ineffective at high temperature.)

At about 400°C a relaxation maximum occurs. It is produced reversibly. It is probably caused by formation, in this temperature range, of para-

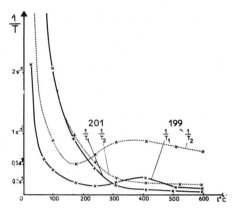

Fig. 1-23. For ^{199}Hg and ^{201}Hg, $1/T_1$ and $1/T_2$ versus temperature; first heating curves (Cagnac and Lemeignan 1967).

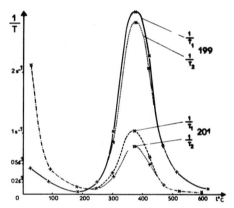

Fig. 1-24. For ^{199}Hg and ^{201}Hg, $1/T_1$ and $1/T_2$ versus temperature after many heating cycles (Cagnac and Lemeignan 1967).

magnetic centers on the wall, centers which disappear again at higher temperature.

Analogous results have been obtained by pumping alkali vapors with their circularly polarized resonance radiation. In this case the orientation obtained is spin orientation of the valence electron. In a glass cell the relaxation times measured are very short, of the order of the time of flight: 10^{-4} sec. The relaxation rate can be reduced considerably by using a paraffin or silicon coating on the wall. For such coatings relaxation times of 10^{-1} and even 1 sec are measured. Mrs. Bouchiat has shown that the relaxation is still 3 to 4 times smaller if deuterated paraffin is used. This result shows that the relaxation process is caused by magnetic interaction between the magnetic moment of the valence electron and the nuclear magnetic moments of the protons or deutons in the wall. For rubidium, Mrs. Bouchiat has studied how the relaxation rate $1/T_1$ depends on the strength of the applied magnetic field H_0.

Figure 1-25 shows the curve obtained for ^{87}Rb, for σ^+ and for σ^- orientation. (The dissymmetry can be explained by the hyperfine inter-

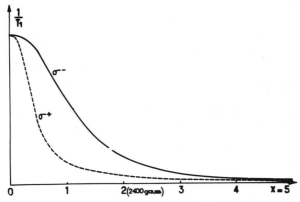

Fig. 1-25. Relaxation of optically oriented Rb atoms on paraffin-coated walls; $\tau_c = 3 \cdot 10^{-10}$ sec (Bouchiat 1963).

action.) With increasing field, the relaxation decreases along a Lorentz curve whose equation can be written:

$$\frac{1}{T} \propto \frac{1}{1 + (\gamma \cdot H_0 \cdot \tau_c)^2}$$

The parameter τ_c introduced has the dimension of time and is interpreted as the "correlation time of relaxation" (the correlation time of the fluctuations of the relaxing stochastic field). For paraffin coatings τ_c is of the order of 10^{-10} sec. For deuterated paraffin the field effect is much smaller (Fig. 1-26). If xenon gas is added to the rubidium vapor, there is a very strong relaxation rate measured in zero field, but the relaxation rate decreases very rapidly if a field is applied. Figure 1-27 shows the Lorentz curve for a xenon pressure of 1.45 Torr. In this case the corresponding correlation time τ_c is about 20 times longer, $\tau_c \approx 2 \cdot 10^{-9}$ sec.

Mrs. Bouchiat and her collaborators could show that this strong relaxation possessing a long correlation time is caused by formation of Rb–Xe molecules. The value of τ_c gives the mean life time of these molecules, which are formed and destroyed by three-body collisions.

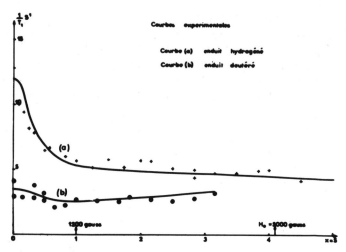

Fig. 1-26. Relaxation of optically oriented Rb atoms on paraffin-coated walls: (a) paraffin coating; (b) deuterated coating (Bouchiat and Brossel 1966).

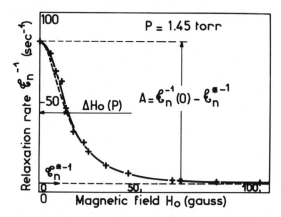

Fig. 1-27. Relaxation rate of oriented Rb versus magnetic field in presence of Kr: $P = 1.45$ Torr (Bouchiat, Bouchiat, and Pottier 1969).

REFERENCES

Althoff, K. 1955. *Z. Physik* 141:33.
Bouchiat, C. A.; Bouchiat, M. A.; and Pottier, L. C. L. 1969. *Phys. Rev.* 181:144.
Bouchiat, M. A. 1963. *J. Physique* 24:611.
Bouchiat, M. A., and Brossel, J. 1966. *Phys. Rev.* 147:41.
Brossel, J. 1952. *Ann. Phys.* Paris 7:622.
Brossel, J., and Bitter, F. 1952. *Phys. Rev.* 86:308.
Bucka, H.; Kopfermann, H.; and Otten, E. W. 1959. *Ann. Physik* 4:39.

Cagnac, B. 1960. *Ann. Phys.* Paris 6:467.
Cagnac, B., and Lemeignan, G. 1967. *Compt. Rend.* Acad. Sci. Paris 264:1850.
Descoubes, J. P.
　1964. *Compt. Rend.* Acad. Sci. Paris 259:327.
　1965. *Compt. Rend.* Acad. Sci. Paris 261:916
　1976. Thesis, Paris.
Pébay-Peyroula, J. C. 1959. *J. de Physique* 20:669.
zu Putlitz, G. 1965. *Ergeb. Exakt. Naturw.* 37:114, 121.

2

ALFRED KASTLER

The Principles of Optical Pumping

PART 2. MAGNETIC RESONANCE AND MAGNETIC MOMENTS

MAGNETIC RESONANCE IN OPTICALLY PUMPED GROUND STATES

We have seen how by optical pumping with circularly polarized light, angular momentum can be transferred to a collection of atoms in the gaseous state and how this gives rise to atomic orientation or to nuclear orientation which is equivalent to creation of a bulk magnetic moment $\mathcal{M}_z$ in the direction of the light beam (Fig. 2-1). In general a magnetic field H_0 is applied in this direction producing a Zeeman splitting between the m levels of the ground state. We have seen also that this orientation process can be monitored by the intensity of the transmitted light $\mathcal{T}_t$ (the vapor becomes more transparent) or by the intensity of the scattered resonance light $\mathcal{T}_s$.

Fig. 2-1. Optical pumping with circularly polarized light (σ^+) creates a magnetic moment in the z direction (longitudinal moment).

Thermal equilibrium is restored by the relaxation process which is essentially caused by wall collisions. We have studied this process in the first chapter and shown that the longitudinal relaxation time depends on the wall temperature and on the value of the magnetic field applied.

When a nonequilibrium situation has been created by optical pumping, population changes restoring thermal equilibrium can also be obtained by magnetic resonance by applying a radiofrequency (RF) field $H_1(\cos \omega t)$. Magnetic dipole transitions are induced if the frequency ω is on resonance with the energy interval $\omega_0 = \Delta E/\hbar$ between energy states of unequal populations. We have already seen an example in the case of excited states: Brossel's experiment.

Magnetic resonance can be achieved also in ground states (between m levels or hyperfine F levels) and can be monitored by a change of the intensity of the light signal ($\mathcal{T}_t$ or $\mathcal{T}_s$).

22

Figure 2-2 reproduces Cagnac's pumping exponential on ^{199}Hg and shows what happens if at time t an RF field at resonance ($\omega = \omega_0$) is suddenly applied. In this transient effect, thermal equilibrium is restored by damped oscillations of the longitudinal magnetic moment $\mathcal{M}_z$. This corresponds to the well-known nutation motion in magnetic resonance. The frequency of nutation is proportional to the amplitude H_1 of the RF field: $\omega_1 = \gamma H_1$.

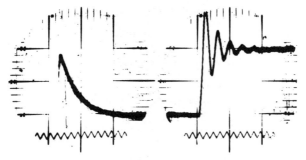

Fig. 2-2. Optical pumping of ^{199}Hg; transient signal of scattered light; sudden application of the radiofrequency field at resonance (Cagnac 1960).

If the time scale is known, ω_1 can be measured, and this measurement gives an absolute value of H_1.

Figure 2-3 shows the nutation process for a high value of H_1. Figure 2-2 and Figure 2-3 are obtained on light scattered by ^{199}Hg. Figure 2-4 shows the nutation process on the light transmitted by pumped Rb. Instead of studying transient signals, the RF field can be permanently applied, and, by tuning the frequency ω or by changing slowly the field

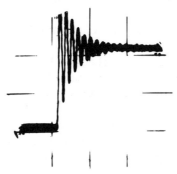

Fig. 2-3. The same transient signal as that in Figure 2-2 with a larger amplitude of the radiofrequency field (Cagnac 1960).

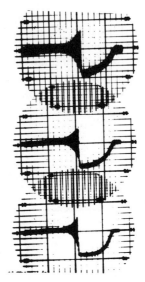

Fig. 2-4. Transient signal of optically pumped Rb; signal of the transmitted light with sudden application of the radiofrequency field (Bouchiat 1963).

H_0, a steady-state resonance curve can be traced. Figure 2-5 shows Cagnac's mounting for ^{199}Hg, and Figure 2-6 shows nuclear magnetic resonance (NMR) curves obtained by Cagnac on ^{199}Hg. Each of the four curves corresponds to a different value of H_1. The half-width of the curve depends on H_1 and also on the pumping-light intensity. By extrapolating to zero field and to zero light intensity, the "natural width" can be obtained, which leads to the value of T_2, the transverse relaxation time of magnetic resonance: $\Delta\omega_0 = 1/T_2$. In the case of ^{199}Hg at room temperature, $\Delta\omega_0/2\pi$ is of the order of 1 hertz, which corresponds to T_2 of the order of 1 sec.

In general T_2 is of the same order of magnitude as T_1 but slightly smaller than T_1. Figure 2-7 shows NMR curves for ^{201}Hg. They are not of Lorentz shape as in the case of ^{199}Hg but of Majorana shape (like Brossel's curve for 6^3P_1). This is always the case if more than two equidistant m levels are present. Figure 2-8 shows resonance curves for ^{113}Cd obtained by Lehmann.

The precise measurement of the center of resonance ($\omega = \omega_0 = \gamma H_0$) leads to a precise determination of γ and of the magnetic moment of the nucleus μ_i in units of the nuclear magneton $\mu_n = \mu_B/1838$:

$$\omega_0 = g_i \frac{\mu_n}{\hbar} H_0 \qquad \frac{\mu_i}{\mu_n} = i g_i$$

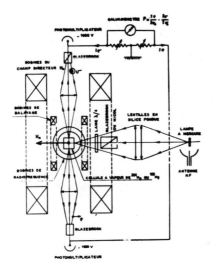

Fig. 2-5. Experimental device for the study of nuclear magnetic resonance of ^{199}Hg and ^{201}Hg (Cagnac 1960).

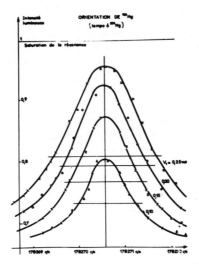

Fig. 2-6. Nuclear magnetic resonance curves of optically pumped ^{199}Hg with a ^{204}Hg lamp (Cagnac 1960).

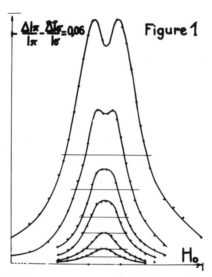

Fig. 2-7. Nuclear magnetic resonance curves of optically pumped ^{201}Hg (Cagnac 1960).

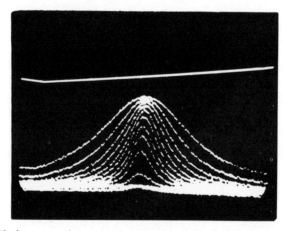

Fig. 2-8. Nuclear magnetic resonance curves of cadmium-113 (Lehmann 1966, 1967).

This needs the precise knowledge of H_0 which is obtained by a conventional proton-resonance measurement in the same field. We give some results to show the precision obtained:

$$\mu_{199} = 0.497865 \pm 6 \cdot 10^{-6} \mu_n \quad \text{and} \quad \mu_{201} = -0.551 \cdot 344 \pm 9 \cdot 10^{-6} \mu_n$$

$$\frac{\mu_{201}}{\mu_{199}} = -1.107\,416\,4 \pm 5 \cdot 10^{-7}$$

The ratio of moments is known with a better precision than the absolute value. It needs no diamagnetic correction and does not depend on the value of the proton moment. For the cadmium isotopes:

$$\frac{\mu_{113}}{\mu_{111}} = 1.046\,084\,0 \pm 2 \cdot 10^{-7}$$

Magnetic resonance has been obtained also in the ground state of optically pumped alkali atoms. The ground state of these atoms is a $^2S_{\frac{1}{2}}$ state, with spherical orbital electron configuration ($L = 0$) and the angular momentum and paramagnetism are caused by the spin $J = S = \frac{1}{2}$ of the valence electron. All alkali nuclei have a nonzero nuclear spin $\mathbf{I}$ which is coupled to $\mathbf{J}$ by the hyperfine coupling $a\mathbf{I} \cdot \mathbf{J}$ and gives rise, in the ground state, to two hyperfine intervals $F_1 = I - \frac{1}{2}$ and $F_2 = I + \frac{1}{2}$ whose interval has been accurately measured by the Rabi method. In a small magnetic field each of the two F states is split into $2F + 1$ magnetic substates m_F corresponding to g factor g_F equal to $g_{F_2} = 2/(2I + 1)$ for the upper and $g_{F_1} = -g_{F_2}$ for the lower F state.

In small fields the linear Zeeman effects give rise to equal absolute values of all Zeeman intervals: $\omega_0 = g_F(e/2m_0)H_0$, but in medium fields (100 to 300 gauss) the decoupling of $\mathbf{I}$ and $\mathbf{J}$ produces the Back–Goudsmit effect. The Zeeman intervals become unequal.

Figure 2-9 shows the energy level diagram of sodium, and Figure 2-10 shows the Zeeman splitting of the two F states of ^{23}Na ($I = \frac{3}{2}$; $F_1 = 1$, $F_2 = 2$). For an RF field of frequency 108.5 MHz four different resonance peaks are expected in the region between 120 to 200 gauss.

Figure 2-11 shows the resonance curves obtained by Cagnac on an optically pumped atomic beam of sodium. At high radiofrequency power, new intermediate and very fine resonance peaks are obtained. They correspond to double-quantum transitions: a jump with two radio-frequency quanta from level m to level $m + 2$. For ^{23}Na three double-quantum transitions are observed, $m_F = -2 \to 0$; $m_F = -1 \to +1$; $m_F = 0 \to +2$. At higher RF powers still higher multiple-quantum transitions can be obtained, as is shown in Figure 2-12 obtained by Barrat on sodium vapor. In multiple-quantum transitions, energy and angular momentum must be conserved: in an n-quantum transition from level m to level $m + n$, n quanta of the same energy and the same polarization (σ^+ quanta each one carrying unit $\hbar$ of angular momentum) contribute to the transition.

These multiple-quantum transitions have been extensively studied by Jacques-Michel Winter in his thesis. He has also predicted and experimentally discovered another type of multiple-quantum transition which

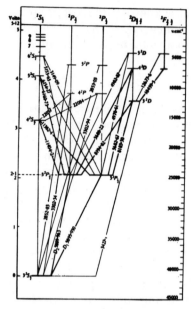

Fig. 2-9. Energy-level diagram of the Na atom.

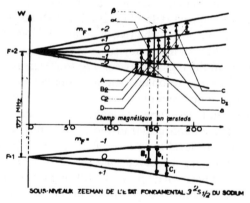

Fig. 2-10. Zeeman sublevels of the ground state of ^{23}Na.

can be produced on a two-level system, $m \rightarrow (m + 1)$, for example $-\frac{1}{2} \rightarrow +\frac{1}{2}$ (Fig. 2-13).

The conservation of energy and of angular momentum shows that a two-quantum transition is produced by a σ^+ and a π quantum. (The latter corresponds to a H_1 component parallel to H_0.) A three-quantum transition is produced by two σ^+ and one σ^- quanta.

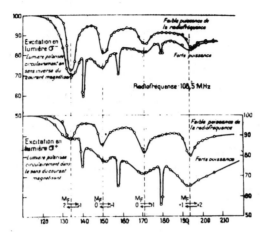

Fig. 2-11. Radiofrequency resonances of the ground state of optically pumped Na atoms in an atomic beam (Brossel, Cagnac, and Kastler 1953).

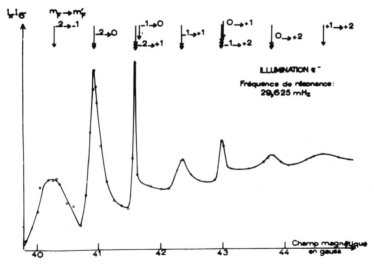

Fig. 2-12. Magnetic resonance of Na ground state in Na vapor; multiple quantum transitions (Barrat, Brossel, and Kastler 1954).

If the RF field H_1 is a rotating field perpendicular to H_1, only the one-quantum transition is possible. If it is an oscillatory field perpendicular to H_0, only odd multiple-quantum transitions are possible (Fig. 2-14). If this oscillatory field is oblique to H_0, all multiple-quantum transitions can occur.

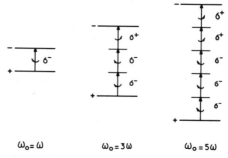

$$\omega_o = \omega \qquad \omega_o = 3\omega \qquad \omega_o = 5\omega$$

Fig. 2-13. Multiple quantum transitions produced on a two-level system.

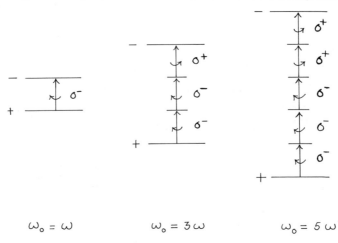

$$\omega_o = \omega \qquad \omega_o = 3\omega \qquad \omega_o = 5\omega$$

Fig. 2-14. J. M. Winter's (1959) multiple quantum transitions.

Figure 2-15 shows transitions of this type observed on ^{23}Na in a very small H_0 field.

Figure 2-16 shows what happens if simultaneously two different radiofrequency fields of frequencies ω and ω' are applied to the atom. A lot of resonances $\omega_0 = p\omega \pm q\omega'$, where p and q are integers, are obtained.

LONGITUDINAL AND TRANSVERSE MAGNETIC MOMENTS

Up to now we have described a collection of atoms in the ground state by the populations of the different m sublevels. Large differences of population can be achieved by optical pumping, and any change in them may be optically detected by monitoring the absorbed or re-emitted light. The production of a nonuniform population distribution

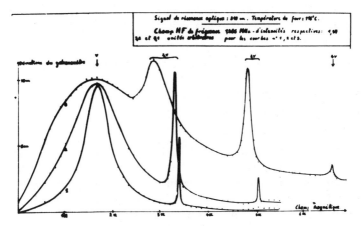

Fig. 2-15. Resonances corresponding to one, two, and three quantum transitions observed on the ground state of sodium; RF field = 1.206 MHz; optical resonance 310 cm; oven temperature 116°C (Margerie and Brossel 1955).

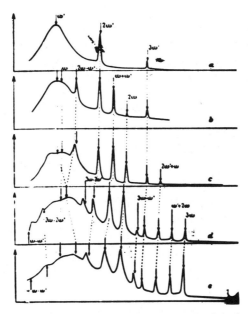

Fig. 2-16. Multiple quantum transitions by simultaneous application of two RF fields ω' and ω to the atom (Brossel, Margerie, Winter 1955).

is related to the creation of a longitudinal bulk magnetic moment $\mathcal{M}_z$ of the vapor (parallel to the magnetic field H_0). We may ask now the following questions:

Is it possible to have a macroscopic transverse magnetic moment perpendicular to H_0?
What is its time behavior in a constant magnetic field H_0?
Can it be monitored by an optical signal?
Can it be produced by optical pumping?

We know that in a magnetic field H_0 an atom in a given energy state undergoes the Larmor precession. Every vectorial property of the atom will precess around the field with the Larmor frequency $\omega_0 = \gamma H_0$. A longitudinal component (like $\mathcal{M}_z$) will not be affected by this precession and will be time independent. A transverse component—as for example a transverse magnetic moment $\mathcal{M}_\perp$—will rotate in the plane xy perpendicular to the field with circular frequency ω_0. The moment $\mathcal{M}_\perp$ of each atom will rotate with a certain phase, and in general the phases of a great number of atoms are distributed at random, and the resultant for a collection of atoms is zero. There is "no coherence" among the atoms. But coherence can be produced by a cause of coherence, for example, by a rotating magnetic field H_1 whose circular frequency ω is equal (or near) the Larmor frequency ω_0. This is precisely the condition of magnetic resonance. The theory of magnetic resonance shows that the rotating field produces a macroscopic rotating magnetic moment among the atoms, and this rotating moment $M_\perp$ is phase locked to the rotating

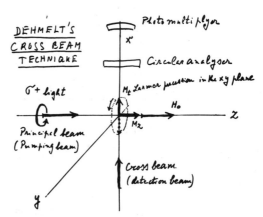

Fig. 2-17. Dehmelt's crossbeam technique: (a) photomultiplier; (b) circular analyzer; (c) principal, pumping, beam; (d) crossbeam, detection beam; M_t is the Larmor precession in the xy plane.

field. This moment has a component parallel to the field H_1 and a component perpendicular to it (in quadrature with it).

We come now to our second question: How can we detect optically the rotating transverse moment? This can be done by the "crossbeam technique" proposed by Dehmelt, illustrated by Figures 2-17 and 2-18. Two light beams are used. They are perpendicular and both circularly polarized (σ^+). The first beam propagates along H_0 (Oz) and produces by optical pumping a longitudinal moment $\mathscr{M}_z$ in the vapor. The second beam, called the crossbeam, is a detection beam crossing the vapor at a right angle to H_0. When a coherent precession of $M_\perp$ is induced, the intensity of this beam is modulated by it at the precession frequency, and this modulation signal can be easily amplified by an a-c technique.

Figure 2-19 shows a transient modulation signal obtained on ^{199}Hg by Cohen-Tannoudji. At time t the pumping beam and the RF field are suddenly cut off. The signal shows the free decay of $\mathscr{M}_\perp$ along an exponential whose time constant is T_2. The crossbeam has to be of small light intensity so that its pumping efficiency can be neglected. Figure 2-20 shows NMR curves of ^{199}Hg obtained by the crossbeam technique. By a phase-sensitive device the components u (in phase with H_1) and v (in quadrature with H_1) can be isolated (Fig. 2-21). Figure 2-22 shows the result obtained for u (dispersion curve) and v (absorption curve).

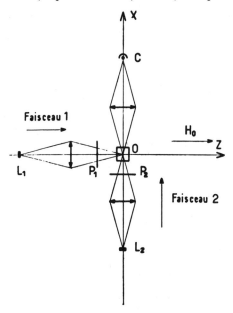

Fig. 2-18. Experimental arrangement for Dehmelt's crossbeam technique.

Fig. 2-19. Dehmelt's crossbeam technique: transient signal of free decay of the transversal moment (Cohen-Tannoudji 1962).

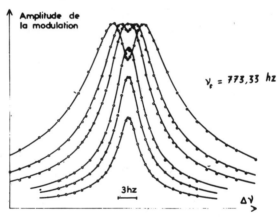

Fig. 2-20. Nuclear magnetic resonance signal observed by the crossbeam technique (Cohen-Tannoudji 1962).

A transverse moment can also be built up directly by optical pumping in a transverse direction ($0x$) as shown in Figure 2-23. This can be done in two ways:

1. If the applied magnetic field H_0 (in the z direction) is high, the pumping has to be made in rhythmic light pulses at the frequency of the Larmor precession in field H_0. The moments induced at each pulse then build up coherently, and a strong macroscopic rotating moment results. This method was suggested by Bell and Bloom.

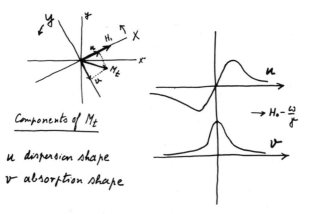

Fig. 2-21. Components of M_t; u = dispersion shape; v = absorption shape.

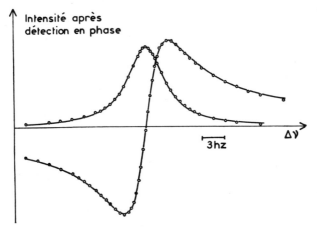

Fig. 2-22. Crossbeam technique with ^{199}Hg; curve of the u component and of the v component of M_t (Cohen-Tannoudji 1962).

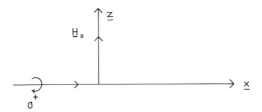

Fig. 2-23. Polarized pumping light beam.

2. If the magnetic field is zero or small, a transverse moment can be produced by steady optical pumping. In this case, we are in the conditions of the Hanle experiment. Figure 2-24 shows the experimental mounting for this experiment. We may discuss the result obtained with the help of Figure 2-25. If $H_0 = 0$, a strong moment M_0 in the $0x$ direction is produced. If now we apply a small field H_0 in the $0z$ direction, this moment will begin to rotate with frequency $\omega_0 = \gamma H_0$ and rotate more and more rapidly if ω_0 is increased. But at the same time this rotating moment is decaying by relaxation. Let us call τ the decay constant. The result will

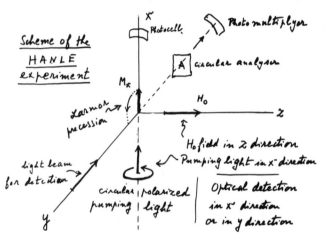

Fig. 2-24. Scheme of the Hanle experiment: (a) light beam for detection; (b) Larmor precession; (c) circular polarized pumping light; (d) pumping light in x^- direction; (e) circular analyzer; (f) photomultiplier; (g) photocell.

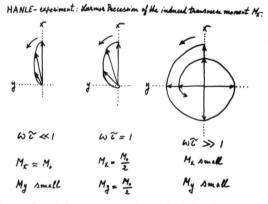

Fig. 2-25. Hanle experiment; Larmor precession of the induced transverse moment M_{x^-}.

be a decaying fan in the $0xy$ plane as is shown in the figure: by the rotation the moment M_x in the x-direction decreases and a moment M_y in the $0y$ direction is produced. The shape of the decaying fan depends on the parameter $\omega_0\tau = \gamma H_0\tau$. Figure 2-25 shows this shape for three cases, $\omega_0\tau \ll 1$, $\omega_0\tau = 1$, and $\omega_0\tau \gg 1$.

As a function of $\omega_0\tau$ the moment M_x is a curve of absorption shape, the moment M_y a curve of dispersion shape. Originally such an experiment was performed by Hanle in 1924 on excited atomic states (for example, Hg 6^3P_1). Figure 2-26 shows the result. Note the scale in H_0 units ($\omega_0\tau = \gamma H_0\tau$). The half-width is of the order 1 gauss. Figure 2-27 shows these two curves obtained on the nuclear ground state of ^{113}Cd (nuclear spin $\frac{1}{2}$). The scale is not indicated, but the half-width in this case is of the order of 10^{-3} gauss. We will see later a case where it is 10^{-6} gauss. The equation of these curves are

$$M_x = \frac{M_0}{1 + (\omega_0\tau)^2} \quad \text{and} \quad M_y = M_0\frac{\omega_0\tau}{1 + (\omega_0\tau)^2}$$

The half-width corresponds to $M_x = M_y = \frac{1}{2}M_0$, $\omega_0\tau = \gamma H_0\tau = 1$. The corresponding value of the field (the critical value) is $H_c = \gamma^{-1}\tau^{-1}$. From the study of the Hanle curve, the decay time τ can be determined if γ is known. Or if τ is known, it can serve to measure $\gamma = g(e/2m_0)$, that is

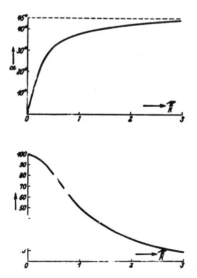

Fig. 2-26. Hanle curve for the excited state 6^3P_1 of the mercury atom: (a) rotation of the plane of polarization; (b) depolarization of resonance radiation (Hanle 1925).

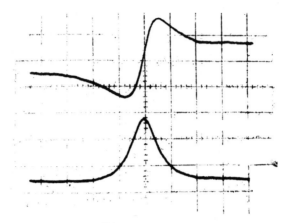

Fig. 2-27. Hanle curves of ^{113}Cd (Lehmann and Cohen-Tannoudji 1964).

to say, the g factor of the state. We will see later an important application of this.

Note that g is proportional to H_c^{-1}. The broader the Hanle curve, the smaller is g.

Level Crossings

The Hanle effect can be considered as a special case of level crossing in atomic spectroscopy (Fig. 2-28). If, in the case of a two-level system (J or $I = \frac{1}{2}$), we plot the energy of the two Zeeman levels $+\frac{1}{2}$ and $-\frac{1}{2}$ as functions of the field H_0, we obtain two straight lines crossing at zero

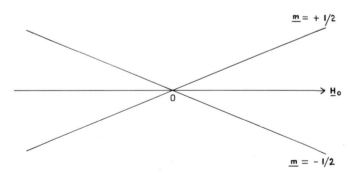

Fig. 2-28. The Hanle effect as a special case of level crossing in atomic spectroscopy.

field. It is at this value of the field that the center of the Hanle curve is located. The half-width of the Hanle curve in the frequency ω_0 scale (or in the energy scale divided by $\hbar$) corresponds to $\omega_0 = \tau^{-1} = \Gamma$.

The term Γ, the reciprocal of the decay time τ, is called the "natural" width. If we study the Hanle effect in an excited state—as Hanle did—we can interpret it as an interference effect at zero field. If we apply a field H we destroy the coherence. Other level crossings can occur between Zeeman levels of a state, especially when fine structure or hyperfine structure is present in the zero field. Figure 2-29 shows such level crossings for state 4^3P of ^4He. There are two crossing points of levels for which $\Delta m = 2$ and three crossing points of levels for which $\Delta m = 1$. If the exciting conditions of the atom are conveniently chosen (coherent excitation), peaks analogous to the Hanle curves can be observed on the crossing points. Such effects were discovered at Michigan University by Peter Franken and his co-workers. Figure 2-30 shows such level-crossing curves (above: dispersion type, below: absorption type) for the two crossing points of 4^3P of ^4He corresponding to $\Delta m = 2$. These curves and many others for excited states of ^4He and ^3He have been obtained by Descoubes using coherent electron-impact excitation. On these experimental curves the field values of the crossing points can be measured and can be calculated from the fine-structure intervals and also (in the case of ^3He) the hyperfine structure intervals. From the half-width of the curves, the life time τ of the state can be obtained.

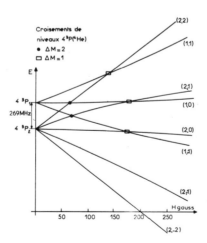

Fig. 2-29. Level crossings in the Zeeman diagram of the fine structure of ^4He (Descoubes 1967).

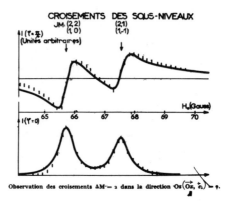

Fig. 2-30. Level crossing signals for ^{4}He; $M = 2$ in the direction $0z$ $(\mathbf{0x}, \mathbf{e}) = \varphi$ (Descoubes 1967).

Table 2-1. ^{4}He and ^{3}He (Descoubes 1967)

Level	Fine structure (MHz)	Hyperfine structure (MHz)	Life time ($\times 10^{-8}$ sec)
$3\,^3$D	$3\,^3D_2 - 3\,^3D_3 = 72.5 \pm 0.5$	$3\,^3D_{3,\frac{5}{2}} - 3\,^3D_{3,\frac{7}{2}} = 55.8 \pm 0.8$	1.7 ± 0.3
$4\,^3$D	$4\,^3D_2 - 4\,^3D_3 = 35.8 \pm 0.4$	$4\,^3D_{3,\frac{5}{2}} - 4\,^3D_{3,\frac{7}{2}} = 28.4 \pm 0.5$	4.2 ± 0.4
$5\,^3$D	$5\,^3D_2 - 5\,^3D_3 = 20.3 \pm 0.3$	$5\,^3D_{3,\frac{5}{2}} - 5\,^3D_{3,\frac{7}{2}} = 15.2 \pm 0.3$	5.9 ± 0.8
$6\,^3$D	$6\,^3D_2 - 6\,^3D_3 = 12.2 \pm 0.3$	$6\,^3D_{3,\frac{5}{2}} - 6\,^3D_{3,\frac{7}{2}} = 9.1 \pm 0.3$	8.2 ± 0.8
$7\,^3$D	$7\,^3D_2 - 7\,^3D_3 = 7.3 \pm 0.3$	$7\,^3D_{3,\frac{5}{2}} - 7\,^3D_{3,\frac{7}{2}} = 4.8 \pm 0.4$	19 ± 7
$5\,^3$F	$5\,^3F_3 - 5\,^3F_4 = 20 \pm 6$		3 ± 2
$3\,^1$D		138 ± 2.5	1.2 ± 0.3
$4\,^1$D		102 ± 1	4.1 ± 0.5
$5\,^1$D		93.8 ± 0.9	4.9 ± 0.5
$6\,^1$D		89.7 ± 0.9	6.1 ± 0.6
$7\,^1$D		118 ± 1.2	7.1 ± 0.8
$3\,^3$P	$3\,^3P_1 - 3\,^3P_2 = 658.55$ $3\,^3P_0 - 3\,^3P_2 = 8300 \pm 400$	$3\,^3P_{2,\frac{3}{2}} - 3\,^3P_{2,\frac{5}{2}} = 540 \pm 1$	8.9 ± 0.5
$4\,^3$P	$4\,^3P_1 - 4\,^3P_2 = 269.0 \pm 0.1$ $4\,^3P_0 - 4\,^3P_2 = 3500 \pm 500$	$4\,^3P_{2,\frac{3}{2}} - 4\,^3P_{2,\frac{5}{2}} = 224.1 \pm 0.2$	16.7 ± 1
$5\,^3$P	$5\,^3P_1 - 5\,^3P_2 = 135.5 \pm 0.1$ $5\,^3P_0 - 5\,^3P_2 = 1550 \pm 300$	$5\,^3P_{2,\frac{3}{2}} - 5\,^3P_{2,\frac{5}{2}} = 113.5 \pm 0.5$	20.7 ± 2
$6\,^3$P	$6\,^3P_1 - 6\,^3P_2 = 77.27 \pm 0.07$ $6\,^3P_0 - 6\,^3P_2 = 1200 \pm 400$	$6\,^3P_{2,\frac{3}{2}} - 6\,^3P_{2,\frac{5}{2}} = 64.0 \pm 0.2$	28 ± 8
$7\,^3$P	$7\,^3P_1 - 7\,^3P_2 = 48.43 \pm 0.15$	$7\,^3P_{2,\frac{3}{2}} - 7\,^3P_{2,\frac{5}{2}} = 38.9 \pm 0.2$	25 ± 8
$8\,^3$P	$8\,^3P_1 - 8\,^3P_2 = 32.49 \pm 0.16$ $8\,^3P_0 - 8\,^3P_2 = 300 \pm 200$	$8\,^3P_{2,\frac{3}{2}} - 8\,^3P_{2,\frac{5}{2}} = 26.1 \pm 0.3$	28 ± 8
$9\,^3$P	$9\,^3P_1 - 9\,^3P_2 = 22.8 \pm 0.8$		

Table 2-1 shows the numerical results obtained by Descoubes for many states of ^{4}He and ^{3}He. Level crossing curves are the result of an interference effect : Consider in an atom a low energy state and in this state Zeeman level m, and in an upper state related to it by a spectral transition, Zeeman levels $m - 1$ and $m + 1$ which are crossing at point P (Fig. 2-31). If we are at a field value far from the crossing points, the two emitted radiations σ^- and σ^+ have different frequencies and are not able to interfere. They are incoherent. If we adjust the field H_0 to be at the crossing point, the two frequencies become equal. They are now coherent and able to interfere, and the result of the interference is a linearly polarized emission **E**.

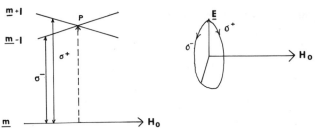

Fig. 2-31. Level crossings are the result of an interference effect.

Transverse Moments, Coherences, and Quantum Mechanics

Let $|E\rangle$ be the different energy eigenstates of an atom. The more general state in which the atom can be described by quantum mechanics is a linear superposition of the $|E\rangle$ states

$$|\psi(t)\rangle = \sum_E a_E(t)|E\rangle$$

All physical predictions concerning the atom in state $|\psi\rangle$ may be expressed in terms of the complex, time-dependent numbers $a_E(t)$, whose time dependance is given by the factor $\exp[-i(E/\hbar)t]$.

For example, the mean value in the state $|\psi\rangle$ of a certain physical grandeur $\mathcal{G}$ (with matrix elements $\mathcal{G}_{EE'}$) is

$$\langle\mathcal{G}\rangle = \sum_{EE'} a_E(t)a_E^*(t)\mathcal{G}_{EE'}$$

Instead of the $a_E(t)$, it is more convenient to introduce the quantities $\sigma_{EE'} = a_E(t)a_{E'}^*(t)$ whose time behavior is given by

$$\exp\left[-i\left(\frac{E - E'}{\hbar}\right)t\right]$$

The σ matrix is the density matrix.

The diagonal elements of this matrix σ_{EE}, $(E = E')$ are time independent; the nondiagonal elements $\sigma_{EE'}$, $(E \neq E')$ are frequency modulated at frequency $\omega = (E - E')/\hbar$, which is the Bohr frequency corresponding to the energy interval.

If the two energy states are magnetic substates E_m and $E_{m'}$ the frequency ω is the Larmor frequency. (There are only nonzero nondiagonal elements if $m' = m \pm 1$).

The term $\langle \mathcal{G} \rangle$ can be expressed in the form:

$$\langle \mathcal{G} \rangle = \sum_{EE'} \mathcal{G}_{EE'} \sigma_{EE'} = \mathrm{Sp} \cdot \sigma \mathcal{G}$$

If $\mathcal{G}$ does not commute with the energy, i.e., if the nondiagonal elements $\mathcal{G}_{EE'}$ are not all zero, $\mathcal{G}$ depends in general on the nondiagonal elements of the density matrix.

Let us now consider an ensemble of N identical atoms. According to statistical mechanics, it can be described by the total density matrix, which is the average of the individual density matrices representing each atom:

$$\rho_{EE'} = \frac{1}{N} \sum_N \sigma_{EE'}$$

$$\langle \mathcal{G}_\rho \rangle = \frac{1}{N} \sum_N \sum_{EE'} \mathcal{G}_{EE'} \sigma_{EE'} = \sum_{EE'} \mathcal{G}_{EE'} \rho_{EE'} = \mathrm{Sp} \cdot \rho \mathcal{G}.$$

The diagonal elements ρ_{EE} of the total density matrix represent the mean probability of the atoms of the ensemble in the state $|E\rangle$. They are the populations of the different energy levels. A longitudinal moment M_z is a grandeur which depends only on these diagonal elements and which, in a steady field H_0, is time independent.

When a nondiagonal element of the total density matrix $\rho_{EE'}$ is different from zero, there is a same cause, common to all atoms, which is responsible for the fact that the relative phase of $a_E(t)$ and $a_{E'}(t)$ does not change in a random way from one atom to another, for example, an RF field or transverse optical pumping. In this case, averaging $\sigma_{EE'} = a_E a_E^*$ over all atoms does not give zero, and the grandeurs $\langle \mathcal{G}_\rho \rangle$ depending on $\rho_{EE'}$ have then a time dependance $\exp -(E - E'/\hbar)t$. In this case, we say that there is a "coherence" between the states $|E\rangle$ and $|E'\rangle$. The nondiagonal element $\rho_{EE'}$ of the total density matrix can be considered by convention as a "measure" of this coherence. Such a physical grandeur, which does not commute with the energy, is a transverse magnetic moment $M_\perp$ in a magnetic field H_0. Its x and y components are time modulated at the Larmor frequency

$$\omega_0 = \frac{E_m - E_{m'}}{\hbar}$$

REFERENCES

Barrat, J. P.; Brossel, J.; and Kastler, A. 1954. *Compt. Rend.* Acad. Sci. Paris 239:1196.
Bouchiat, M. A. 1963. *J. Physique* 24:611.
Brossel, J.; Cagnac, B.; and Kastler, A. 1953. *Compt. Rend.* Acad. Sci. Paris 237:984.
Brossel, J.; Margerie, J.; and Winter, J. M. 1955. *Compt. Rend.* Acad. Sci. Paris 241:555.
Cagnac, B. 1960. *Ann. Phys.* Paris 6:467.
Cohen-Tannoudji, C. 1962. *Ann. Phys.* Paris 7:423, 469.
Descoubes, J. P. 1967. Thesis, Paris.
Franken, P. A. 1961. *Phys. Rev.* 121:508.
Hanle, W. 1925. *Ergeb. Exakt. Naturw.* 4:214.
Lehmann, J. C.
   1966. Thesis, Paris.
   1967. *Ann. Phys.* Paris 2:345.
Lehmann, J. C., and Cohen-Tannoudji, C. 1964. *Compt. Rend.* Acad. Sci. Paris 258:4463.
Margerie, J., and Brossel, J. 1955. *Compt. Rend.* Acad. Sci. Paris 241:373.
Winter, J. M. 1959. *Ann. Phys.* Paris 4:745.

3

ALFRED KASTLER

Virtual Transitions in Atomic Spectroscopy

DEFINITION OF REAL AND VIRTUAL TRANSITIONS

Photons can be absorbed by an atom only in a given quantum state of energy E_1 if their frequency ω satisfies the Bohr condition

$$\hbar\omega = E_2 - E_1$$

where E_2 is the energy of a higher quantum state. This condition is the expression of conservation of energy. The energy of the light quantum is transferred to the atom. In the reverse process of photon emission by an atom, the Bohr condition is also satisfied. But in such processes of energy transfer between the atom and the electromagnetic radiation field a second condition must also be fulfilled, the condition of conservation of angular momentum. It determines the state of polarization of the radiation which interacts with the atom, π for a $\Delta m = 0$ transition, σ^+ for a $\Delta m = +1$ transition, σ^- for a $\Delta m = -1$ transition. We say that such transitions which lead to transfer of energy and of angular momentum between the atom and the radiation field are "real transitions." In an ensemble of atoms such real transitions produce changes of population of the concerned energy states. They change the diagonal elements of the density matrix.

What happens (Fig. 3-1) if the atom is in the presence of a radiation field containing photons whose energy does not correspond to the Bohr condition $\hbar\omega' \neq E_2 - E_1$ or whose polarization is not the right one to satisfy the condition of transfer of angular momentum?

These photons cannot be absorbed by the atom, and no population changes in the ensemble of atoms can be produced. Nevertheless there is a small interaction between the atoms and the radiation field. We know that the propagation of the light is changed by the presence of non-absorbing matter. It is the phenomenon of "dispersion." Something must be changed also, during this interaction, in the properties of the atoms. What? Let us first ask the question: Why is such an interaction possible in spite of the energy discrepancy δE?

44

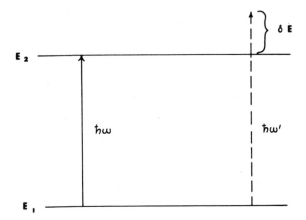

Fig. 3-1. Interaction among atoms in presence of radiation field with no population change.

According to Heisenberg's fourth uncertainty relation

$$\delta E \times \delta t \approx \hbar$$

we know that if the duration δt of a process is very small, the energy E corresponding to this process is defined only with an indetermination of the order of $\delta E \approx \hbar/\delta t$. Reciprocally, if for a given process there is an energy discrepancy δE, the process is possible if the time of duration of this process is no longer than

$$\delta t \approx \frac{\delta E}{\hbar}$$

The uncertainty relation makes possible an interaction of an atom with photons of any frequency (or any state of polarization) provided the time of interaction is very short, short enough to satisfy the uncertainty relation.

Such short-time interactions are called "virtual transitions." Such transitions do not change the diagonal elements of the density matrix of an ensemble of atoms (the populations), but they may change the nondiagonal elements (the coherences). We can describe such virtual transitions by Feynman diagrams (Fig. 3-2 and Fig. 3-3).

An atom can virtually emit photons of any frequency if it absorbs them very rapidly again or, if photons of any frequency are present in its neighborhood, it can absorb them virtually if it emits them very rapidly again. We may now investigate in detail some examples of such virtual transition processes.

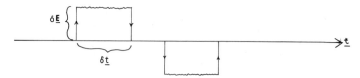

Fig. 3-2. Virtual emission and reabsorption of photons by an atom.

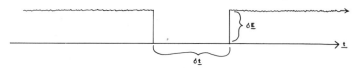

Fig. 3-3. Virtual absorption and reemission of photons by an atom.

VIRTUAL TRANSITIONS PRODUCED BY A RESONANT RF FIELD

Figure 3-4 shows three different cases of optical pumping arrangements, which we may distinguish by the letters a, b, and c:

In all three cases we apply to the atoms a steady magnetic field H_0; a radiofrequency field $H_1 \cos \omega t$; and a pumping light beam F which is circularly polarized. But the spatial arrangement of these elements is different in each case.

In case a the RF field is perpendicular to the steady field H_0, and the optical pumping is operating in the longitudinal direction, the light beam F being parallel to H_0; in case b the RF field is also perpendicular to H_0, but the optical pumping is in the transverse direction, perpendicular to H_0; finally, in case c, the optical pumping is also in the transverse direction, but the RF field is parallel to H_0.

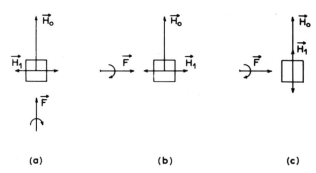

Fig. 3-4. Three different cases of optical pumping: (a) real transitions; (b) virtual transitions; and (c) virtual transitions; H_0 is a steady magnetic field; H_1 is an oscillating radiofrequency field; and F is a circularly polarized light beam.

We may analyze successively the three cases for a simple atomic system having only two Zeeman levels (^{199}Hg: $m_i = +\frac{1}{2}$ and $m_i = -\frac{1}{2}$).

Case a: In case a the optical pumping produces in the ensemble of atoms a longitudinal magnetic moment M_z. It changes the populations of the Zeemen levels. The radiofrequency field $H_1 \cos \omega t$ produces real transitions if the resonance condition, $\omega_0 = (2p + 1)\omega$, is fulfilled, p being an integer. We have already seen that in these transitions energy and angular momentum are conserved. Figure 3-5 shows these transitions, as they are experimentally observed on the ground state of ^{199}Hg for different intensities of the RF field H_1 by Cohen-Tannoudji. We note that, if the amplitude is enhanced, these resonances are displaced and are broadened.

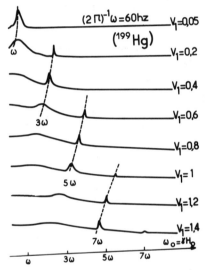

Fig. 3-5. Case a: Real transitions obtained on the ground state of ^{199}Hg ($i = \frac{1}{2}$) and $(2\Gamma)^{-1}\omega = 60$ Hz (Cohen-Tannoudji 1968).

In this case of real transitions, the optical pumping process and the RF resonances change the diagonal elements of the total density matrix of the atoms.

Case b: In case b the optical pumping operates in a direction perpendicular to H_0 and produces a transverse magnetic moment M_t in the ensemble of atoms, or it would produce such a moment if the field H_0 would be small. (Remember the Hanle experiment.) But now we are in high-field conditions ($\omega_0 \tau \gg 1$), and this moment will be destroyed by the Larmor precession if no RF field is present.

Theory shows that if we now apply a RF field $H_1 \cos \omega t$ to the atoms and if the frequency of this field satisfies the resonance condition, $\omega_0 = 2p\omega$ where p is an integer, then coherence is induced in the Larmor precession of the ensemble of atoms, and this coherence is monitored by the intensity of the transmitted F beam, measured by a photodetector. Figure 3-6 shows a suite of such resonances monitored by the light signal. Each of these resonances has been traced with a different value of the H_1 amplitude. (The voltage V applied to the RF coil is proportional to H_1.) These resonances become stronger if H_1 is enhanced and they are displaced, but they are *not* broadened. These resonances, where an even number of photons interact with the atom, correspond typically to virtual transitions. There is no change of the diagonal elements of the density matrix, but there is a creation of a nonzero nondiagonal element of this matrix (creation of coherence).

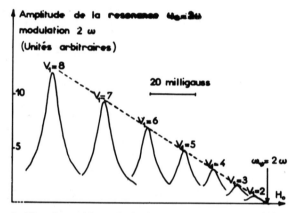

Fig. 3-6. Case b: Virtual transitions obtained on the ground state of ^{199}Hg for $\omega_0 = 2\omega$ (Cohen-Tannoudji and Haroche 1965).

Resonances of this type have been studied in detail by Cohen-Tannoudji and Haroche. Figure 3-6 shows a series of resonances for $\omega_0 = 2\omega$. Figure 3-7 shows the resonances for $\omega_0 = 4\omega$. Figure 3-8 shows the displacements in function of H_1^2 for $\omega_0 = 2\omega$, $\omega_0 = 4\omega$, and $\omega_0 = 6\omega$. Note that the straight lines are the theoretical predictions, and the points are the experimental results.

Case c: We come now to case c. Here also, as in the preceding case, the optical pumping is transverse pumping and produces no population changes. No real transitions can be induced in this case. The RF field is parallel to the H_0 field. Classically this represents a "frequency-modulated" field.

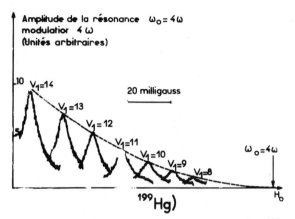

Fig. 3-7. Case b: Virtual transitions obtained on the ground state of ^{199}Hg for $\omega_0 = 4\omega$ (Cohen-Tannoudji and Haroche 1965).

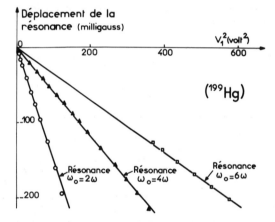

Fig. 3-8. Displacements of the case b resonances in ^{199}Hg as a function of the amplitude H_1 of the radiofrequency field applied for ^{199}Hg (Cohen-Tannoudji and Haroche 1965).

Theory shows that in this case virtual transitions giving rise to optical signals on the F beam are produced, if the resonance condition $\omega_0 = p\omega$ is satisfied, p being an integer (even or odd in this case).

Figure 3-9 shows such a resonance for $p = 1$. In this case, if the amplitude H_1 is increased, the resonances are neither broadened nor displaced. Figure 3-10 shows that their half-width is independent of H_1, $\omega_1 = \gamma H_1$.

Resonances of this type were first observed on excited states of atoms (cadmium) by Alexandrov in Leningrad and by Geneux in Switzerland. On the ground state of ^{199}Hg these resonances have been theoretically

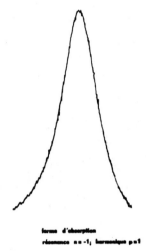

forme d'absorption
résonance n = -1; harmonique p = 1

Fig. 3-9. Case c resonance; ^{199}Hg; $\omega/2\pi = 770$ Hz; form of absorption, resonance $n = 1$; harmonic $p = 1$ (Polonsky 1966).

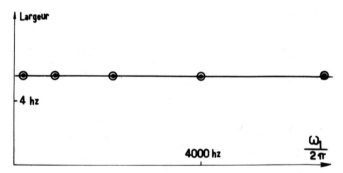

Fig. 3-10. This figure shows that the width of the case c resonance is independent of the amplitude H_1 of the applied radiofrequency field, $\omega_1 = \gamma H_1$ (Polonsky and Cohen-Tannoudji 1965).

and experimentally investigated by Cohen-Tannoudji and Miss Polonsky. These resonances can be detected not only on the fundamental frequency ω; they give rise also to optical signals on the harmonics $q\omega$. A specially interesting case is the case $p = 0$, $q = 1$. It means that we observe near the field value $H_0 = 0$, that we apply a radiofrequency field $H_1 \cos \omega t$, and that we observe an a-c signal at frequency ω. This signal has dispersion shape. It is shown by Figure 4-14. We may call it the "frequency-modulated" Hanle experiment.

In conclusion, in cases b and c the simultaneous action of the transverse optical pumping and of the RF field induces coherences in the ensemble of atoms, and these coherences can be monitored by the intensity modulation of a transverse light beam, either the pumping beam itself or another independent light beam that is propagating in a transverse direction.

VIRTUAL TRANSITIONS PRODUCED BY A NONRESONANT RF FIELD

In the preceding cases we have studied the effect of virtual transitions produced by a RF field. To observe these effects the frequency of the RF field has to satisfy a resonance condition, $\omega_0 = p\omega$ or $\omega_0 = 2p\omega$. The energy conservation would have been satisfied for obtaining real transitions, but the conservation of angular momentum would not have been satisfied.

We will study now an example of virtual transitions which can be obtained with RF radiation of any frequency ω. Illustrating the virtual transitions by Feynman diagrams, we have noted that an atomic system is able to emit virtually and reabsorb photons of any frequency if this process takes place in a time so short that the Heisenberg relation is satisfied. Even a free electron can do so, and Schwinger has pointed out that this virtual emission process should modify slightly the g factor of the electron, that this factor should not be exactly 2, but $2(1 + \varepsilon)$ where ε is of the order 10^{-3}. This effect has been experimentally confirmed by Kusch and Foley, and since ε has been measured with great accuracy by Crane's bottle experiment.

Such spontaneous virtual emission processes of an atomic system occur if this system is in vacuum and the magnitude of the effect is imposed by nature.

If the atom is surrounded by a radiation field then the reverse process of induced virtual absorption followed by emission must occur, as was first pointed out by Cohen-Tannoudji, and the magnitude of this process will depend on the energy density of the field surrounding the atom and on the energy of its quanta (the frequency). Such an atom surrounded by a radiation field is called by Cohen-Tannoudji a "dressed atom."

The virtual absorptions produced by the radiation field will change the g factor of the atom. This effect, predicted by Cohen-Tannoudji, has been experimentally established by Haroche on ^{199}Hg in the following way:

We have seen that in the Hanle experiment, the half-width of the Hanle curve in the H_0 scale is

$$H_c = \gamma^{-1}\tau^{-1} \qquad \text{where} \qquad \gamma = g\frac{e}{2m_0}$$

If g varies, τ being kept constant, the half-width will vary proportionally to $H_c \propto g^{-1}$; the smaller the g factor, the larger the half-width. A first Hanle experiment is made in normal conditions: H_0 in the $0z$ direction, the σ^+ optical pumping beam in the $0x$ direction.

Figure 3-11 shows the normal Hanle curve of the nuclear Zeeman splitting of ^{199}Hg. It is curve $V_1 = 0$. Then a RF field $H_1 \cos \omega t$ is applied in the $0y$ direction, and a new Hanle curve is traced. The curves $V_1 = 15$, $V_1 = 17.5$, etc., correspond to increasing values of the amplitude H_1 of the RF field, hence to decreasing values of the g factor. For $V_1 = 19$, the Hanle curve has become infinitely large; the g factor of the atom is zero. For this H_1 value, the presence of the RF field suppresses the Zeeman splitting of the H_0 field. For higher values of the RF field, the Hanle curves become narrower, but the sign of g is reversed.

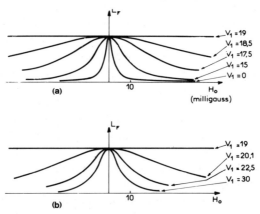

Fig. 3-11. Hanle curves of ^{199}Hg ground state in the presence of a radiofrequency field $H_1 \cos \omega t$. At (a) the V_1 voltage applied to the RF coil is proportional to H_1; at (b) the sign of g is reversed (Cohen-Tannoudji and Haroche 1966).

Figure 3-12 shows how the g factor depends on H_1. On the abscissa is plotted the dimensionless variable ω_1/ω, where $\omega_1 = \gamma H_1$, on the ordinate, the ratio of g to its normal value g_0. The curve obtained is a J_0 Bessel function.

VIRTUAL TRANSITIONS PRODUCED BY AN OPTICAL RADIATION FIELD

Virtual emission and absorption processes can also occur with photons of the optical frequency range. They shift slightly the energy of atomic quantum states. The Lamb shift is a result of virtual emission and re-

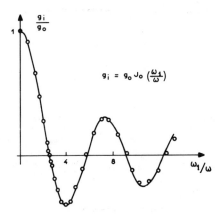

$$g_i = g_0 J_0\left(\frac{\omega_1}{\omega}\right)$$

Fig. 3-12. Change of the g factor of an atom by a radiofrequency field, $H_1 \cos \omega t$, for the ^{199}Hg ground state; $g_i = g_0 J_0(\omega_1/\omega)$ Bessel function where $\omega_1 = \gamma H_1$ (Cohen-Tannoudji and Haroche 1966).

absorption of photons by the atom in vacuum. If the atom is a "dressed atom," if it is surrounded by light radiation emitted by a lamp, virtual absorption processes also shift the energy levels. We may call this effect the lamp shift; it was predicted by Barrat and Cohen-Tannoudji and confirmed experimentally by the latter in his thesis. This effect is intimately connected to the phenomena of optical dispersion.

Consider Figures 3-13 and 3-14 relative to ^{199}Hg and its states 6^1S_0 and 6^3P_1. Photons of wave number k_0 can be absorbed by the atom. They cause real transitions and produce optical resonance. Photons of different wave number, $k \neq k_0$, outside the absorption band of the atom, are not absorbed by the atom. They are dispersed by it, strongly if they are in the anomalous dispersion range. These photons are virtually absorbed and reemitted by the atom and produce shifts of the energy

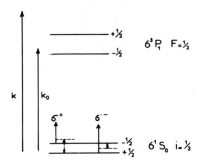

Fig. 3-13. Principle of the light-shift experiment for ^{199}Hg.

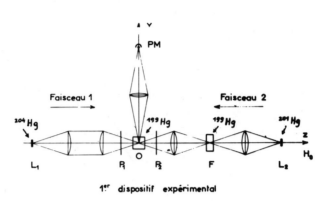

Fig. 3-14. Experimental arrangement for the light shift on ^{199}Hg ground state (Cohen-Tannoudji 1962).

levels. These small shifts can be detected by looking on the NMR transition of the ground state and by measuring its frequency interval ω_0 in the absence and in the presence of the shifting k radiation. If this k radiation is σ^+ polarized, it will act only on level $m_i = -\frac{1}{2}$ and shift it upward. (The sign of the shift depends on the sign of the wave number difference $k - k_0$.) The frequency interval ω_0 will become larger. If the k radiation is σ^- polarized, it will act only on level $m_i = +\frac{1}{2}$ and shift it upward also. The frequency interval ω_0 will become smaller.

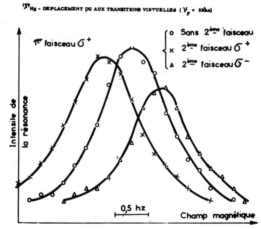

Fig. 3-15. Displacement caused by virtual transitions in ^{199}Hg, $v_f = 4\,\text{kHz}$; $\bigcirc$ is the curve without a second beam; $\times$ is the curve with second beam σ^+; and $\triangle$ is the curve with second beam σ^- (Cohen-Tannoudji 1962).

Figure 3-15 shows these shifting effects. The shifting light k comes from a lamp containing ^{201}Hg which emits a hyperfine component a which is near, but not coincident with, the A component of ^{199}Hg. The wave number difference in this case is of the order of the Doppler breadth of the lines (Fig. 3-16). The two lines are slightly overlapping, but a filter of ^{199}Hg vapor inserted between the lamp and the resonance cell suppresses the absorption effect. The shifts observed are of the order of 0.5 hertz.

Figure 3-17 shows that this lamp shift depends linearly on the intensity of the shifting light beam.

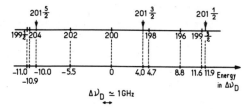

Hyperfine structure of $\lambda = 2537\,\text{Å}$ line

Fig. 3-16. Hyperfine structure of $\lambda = 2537$ Å line.

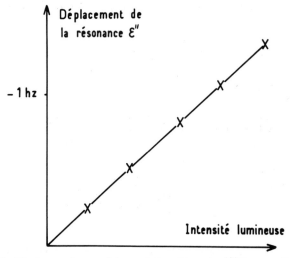

^{199}Hg - DÉPLACEMENT DÛ AUX TRANSITIONS RÉELLES.
$(\mathcal{V}_f = 773, 33\text{ hz})$

Fig. 3-17. Displacement caused by real transitions in ^{199}Hg, $\nu_i = 773.33$ Hz (Cohen-Tannoudji 1962).

The wave number k of the shifting light can be tuned by magnetic scanning.

Figure 3-18 shows how the shift depends on $k - k_0$. The curve looks like an anomalous dispersion curve. This is not surprising. This shifting effect is the reciprocal of anomalous dispersion. Dispersion is the effect of the atoms on the light. The lamp shift is the effect of the light on the atoms.

Figure 3-19 shows how Cohen-Tannoudji and Haroche succeeded in enhancing this shifting effect. They used as a shifting lamp a lamp of

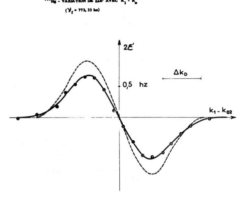

Fig. 3-18. Light shift as a function of $k - k_0$ in ^{199}Hg with $v_f = 773.33$ Hz (Cohen-Tannoudji 1962).

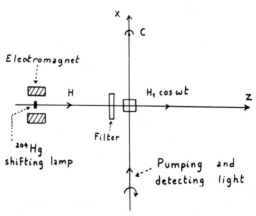

Fig. 3-19. Experimental arrangement for enhancing the light-shift effect: (a) electromagnet; (b) ^{204}Hg shifting lamp; (c) filter; (d) pumping and detecting light; $H_1 \cos \omega t$ (Dupont-Roc, Polonsky, Cohen-Tannoudji, and Kastler 1967a).

^{204}Hg in a strong longitudinal magnetic field. This lamp emits two circularly polarized Zeeman components, σ^+ for which $k - k_0 > 0$ and σ^- for which $k - k_0 < 0$. The shiftings produced by both add their effects. In this way a shift $\Delta\omega_0$ of 6 hertz is obtained (Fig. 3-20). If the ^{199}Hg atoms of the resonance cell are in a zero magnetic field ($H_0 = 0$), the two levels $m_i = \pm\frac{1}{2}$ are coincident. In this case the shifting light produces a removal of degeneracy, a splitting of the level quite analogous to a Zeeman splitting. This splitting must be of 6 hertz. This has been directly demonstrated by the following experiment. Let us return to Figure 3-19 and insert a shutter, stopping the shifting light. The pumping beam will build up a magnetic moment M_x in the x direction. The shutter is suddenly removed, and the shifting beam is then equivalent to a small steady magnetic field (which we may call a "fictitious" field, H_f) in the z direction. This suddenly applied H_f field will produce the Larmor precession of M in the xy plane and will cause the modulation of the intensity of the pumping beam.

Figure 3-21 shows this modulation effect. The Larmor frequency measured on this slide is 6 hertz.

Figure 3-22 shows the combined effect of the fictitious field H_f and of a real applied field H_0. If both fields are parallel, the resultant field is $H_f + H_0$. The linear Zeeman line results. But if both fields are perpendicular, the resultant field is $(H_f^2 + H_0^2)^{\frac{1}{2}}$, and the hyperbolic curve results.

Analogous results have been obtained on ^{201}Hg (Fig. 3-23). For this isotope, the nuclear spin is $i = \frac{3}{2}$, and in a magnetic field H_0 the ground state is split in four equidistant nuclear m states ranging from $m_i = -\frac{3}{2}$

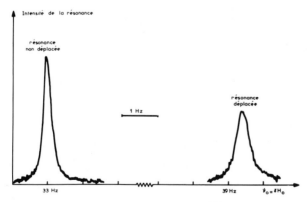

Fig. 3-20. Light shift on nuclear magnetic resonance of ^{199}Hg; (a) nondisplaced line; (b) displaced line.

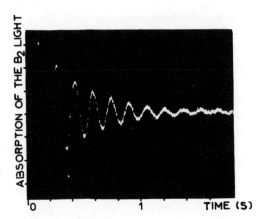

Fig. 3-21. Larmor precession of the ^{199}Hg nuclear spins in the fictitious field associated with a light beam B_1 detected on the absorption of light beam B_2 (Dupont-Roc, Polonsky, Cohen-Tannoudji, and Kastler 1967b).

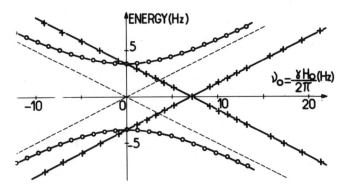

Fig. 3-22. Zeeman diagram of the ground state of ^{199}Hg perturbed by a nonresonant light beam: dashed lines show normal Zeeman effect; + show light beam and magnetic field parallel; ○ show light beam and magnetic field perpendicular (Dupont-Roc, Polonsky, Cohen-Tannoudji, and Kastler 1967b).

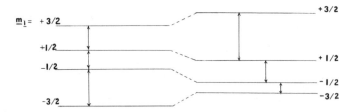

Fig. 3-23. Effect of a linearly polarized shifting light beam in producing a fictitious electric field in the ground state of ^{201}Hg.

to $m_i = +\frac{3}{2}$. If a RF field $H_1 \cos \omega t$ is applied, only one NMR frequency is observed. If now we apply a shifting light beam which is linearly polarized with its **E** vector parallel to H_0, its effect is equivalent to that of a fictitious electric field.

A Stark effect results and the three intervals become unequal. Figure 3-24 shows the experimental arrangement. The shifting light is produced by a light source of ^{200}Hg.

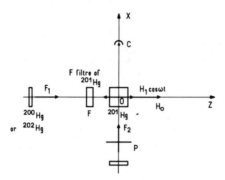

Fig. 3-24. Experimental arrangement for the study of light shifts on the ground state of ^{201}Hg; $i = \frac{3}{2}$.

Figure 3-25 shows the Stark splitting of the NMR line (Cagnac).

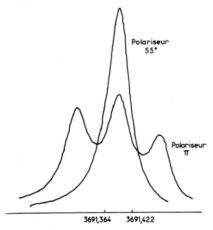

Fig. 3-25. Splitting of the nuclear magnetic resonance curve of the ground state of ^{201}Hg by a light beam of a ^{202}Hg light source linearly polarized at 55° to the steady magnetic field (Cagnac, Israël, and Nogaret 1968).

Figure 3-26 shows the evolution of the four m_i levels with field H_0 in the presence of the fictitious Stark splitting. Instead of one level crossing at zero field—in the absence of this splitting—we have now four nonzero-level crossings.

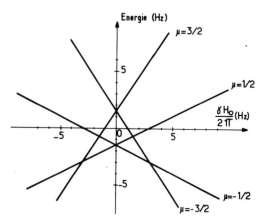

Fig. 3-26. Zeeman levels of the ground state of ^{201}Hg, the light shift producing a Stark effect. Note the four nonzero level crossings ⊗.

Figure 3-27 shows the level-crossing signals observed by Cohen-Tannoudji and Dupont-Roc.

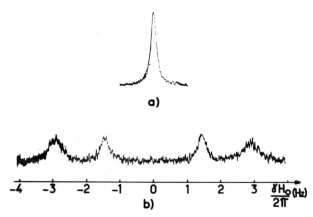

Fig. 3-27. Level-crossing signals in the ground state of ^{201}Hg: (a) without light shift, the zero-level crossing; (b) with light shift, the four nonzero-level crossings (Dupont-Roc and Cohen-Tannoudji 1968).

CONCLUDING REMARKS

Virtual transitions are exchange processes between atoms and radiation fields which do not affect the diagonal elements of the density matrix (they do not produce population changes), but they create nondiagonal elements of the density matrix in an ensemble of atoms (coherences). They can produce on the atoms important effects:

Virtual spontaneous emission in vacuum changes slightly the g factor of the electron and produces the Lamb shift of atomic quantum states.

Induced virtual absorption in dressed atoms can also change g factors and shift energy levels.

We have seen a case where—by the presence of a RF field— the Zeeman splitting in a steady field is suppressed, and we have seen another case where, in the absence of any magnetic field, a Zeeman splitting can be produced by the application of a light beam.

The effects described have been obtained with conventional light sources. With laser light, these effects can be amplified by many orders of magnitude.

REFERENCES

Cagnac, B.; Israël, A.; and Nogaret, M. 1968. *Compt. Rend.* Acad. Sci. Paris 267:274.
Cohen-Tannoudji, C.
   1962. *Ann. Phys.* Paris 7:423, 469.
   1968. *Cargèse Lectures in Physics*, vol. II. New York: Gordon and Breach.
Cohen-Tannoudji, C., and Haroche, S.
   1965. *Compt. Rend.* Acad. Sci. Paris 261:5400.
   1966. *Compt. Rend.* Acad. Sci. Paris 262:268B.
Dupont-Roc, J., and Cohen-Tannoudji, C. 1968. *Compt. Rend.* Acad. Sci. Paris 267:1211.
Dupont-Roc, J.; Polonsky, N.; Cohen-Tannoudji, C.; and Kastler, A.
   1967a. *Compt. Rend.* Acad. Sci. Paris 264:1811.
   1967b. *Phys. Letters* 25A:87.
Polonsky, N. 1966. Thesis, 3d cycle, Paris.
Polonsky, N., and Cohen-Tannoudji, C. 1965. *Compt. Rend.* Acad. Sci. Paris 260:5231.

4

ALFRED KASTLER

The Franck–Condon Principle and Very Small Magnetic Fields

THE FRANCK–CONDON PRINCIPLE IN ATOMIC SPECTROSCOPY

We have seen, on the example of ^{199}Hg, how nuclear orientation can be achieved by optical pumping. To be efficient, three conditions have to be fulfilled: the spectral transition used for pumping must have a high transition probability, the light source used must produce an intense, not self-reversed spectral line, the relaxation process which counteracts the pumping process must have a small speed. These conditions are fulfilled for the Hg line 2537 Å, $6^1S_0 \rightarrow 6^3P_1$. They are still better fulfilled for the Zn line 2139 Å, $4^1S_0 \rightarrow 6^1P_1$, which has a very strong transition probability. The scheme of optical pumping for the odd ^{67}Zn isotope ($i = \frac{1}{2}$) is the same as for ^{199}Hg. We asked Lehmann to achieve

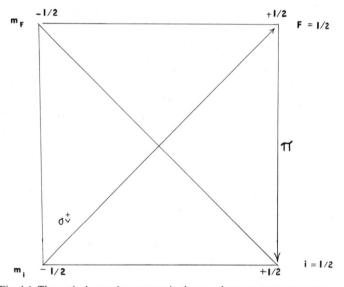

Fig. 4-1. The optical pumping process in the m_F scheme as a two-step process.

62

nuclear orientation of ^{67}Zn by optical pumping. The experiment failed. Why did it fail? We had to explain it.

If we look on the optical pumping process in the m_F scheme, it seems to be a two-step process (Fig. 4-1): Absorption of a σ^+ photon leading from the $m_i = -\frac{1}{2}$ level of the ground state to the $m_F = +\frac{1}{2}$ level of the excited state, followed by spontaneous emission leading down to the ground state.

A thorough analysis of the pumping process shows that it is in fact a three-step process: (a) a spectral transition of absorption of electromagnetic energy; (b) the evolution of the atom in the upper state during life time τ; and (c) a spectral transition of emission of electromagnetic energy.

The nuclear orientation process takes place only during step b. There is no change of nuclear orientation during steps a and c. We may consider this as a consequence of the Franck–Condon principle. This principle was developed for molecular spectroscopy and can be expressed by the following statement:

> In a sudden process which affects the electron configuration (a spectral transition or a molecular collision), the positions of the nuclei remain unchanged.

In atomic spectroscopy, we are not interested in the position of the nuclei, we are interested in the direction of the nuclear axis, and we can make the corresponding statement:

> In a sudden process which affects the electron configuration (a spectral transition or a molecular collision), the orientation of the nuclear axis remains unchanged.

According to this principle steps a and c of the optical pumping process, which are spectral transitions, do not contribute to the change of orientation of the nucleus. This cannot be seen on the m_F scheme. We have to go back to the m_J, m_i scheme as illustrated in Figure 4-2.

In this scheme the m_i states are separated vertically, the m_J states horizontally. We start from level $m_J = 0$, $m_i = -\frac{1}{2}$ in the ground state $^1S_0$ and we reach, by σ^+ absorption, the level $m_J = +1$, $m_i = -\frac{1}{2}$ in the upper state P_1. According to the Franck-Condon principle in the spectral transition, m_i remains unchanged, only m_J changes. If nothing happens in the upper state, the only way down is the return to the original level by σ^+ emission.

But during the stay of the atom in the upper state, the hyperfine coupling $a\mathbf{i}\mathbf{J}$ sets in, and the vectors $\mathbf{i}$ and $\mathbf{J}$ precess around their resultant vector

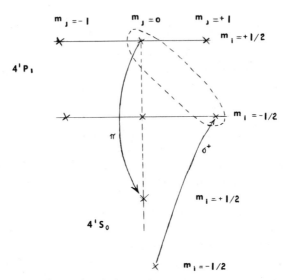

Fig. 4-2. The m_J, m_i scheme of optical pumping to show spectral transitions and nuclear orientation.

F (Fig. 4-3). It is this precession which produces the change of orientation of the nuclear axis. This precession mixes the two levels $m_J = +1$, $m_i = -\frac{1}{2}$ and $m_J = 0$, $m_i = +\frac{1}{2}$ (which have the same m_F value) of the upper state. This is indicated in our scheme by the dotted closed curve. We see that from level $m_J = 0$, $m_i = +\frac{1}{2}$ the atom can fall down by

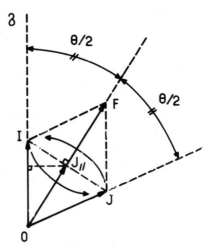

Fig. 4-3. The vector model for hyperfine coupling, $\mathbf{J} + \mathbf{I} = \mathbf{F}$.

emission of a π photon to level $m_J = 0$, $m_i = +\frac{1}{2}$ of the ground state. Again in this spectral transition m_i does not change, but in its final state the nucleus has the orientation $m_i = +\frac{1}{2}$. The nuclear spin has been reversed, and this has been achieved during step b.

This analysis shows what must be the condition that the precession around **F** changes the orientation of the **i** vector. The time available for this precession is the lifetime τ of the atom in the upper state. The angular velocity of precession is given by the coupling constant a in the frequency scale and the mean angle θ of precession of an atom is of the order of $a\tau$. To produce a significant change of the direction of **i** this angle must be at least of the order of 1 radian:

$$a\tau \geqslant 1 \text{ radian} \qquad \text{or} \qquad a \geqslant \Gamma$$

If we introduce the "natural width" of the upper state $\Gamma = \tau^{-1}$. Condition $a \geqslant \Gamma$ states that the hyperfine structure of the level a must be larger or at least as large as the natural line width Γ. This is the case for ^{199}Hg, where the hyperfine structure is widely resolved ($a \approx 22.000$ MHz, $\Gamma \approx 1$ MHz). It is not the case for level 4^1P_1 of ^{67}Zn, where the hyperfine structure (HFS) is small and not resolved ($\Gamma \approx 100$ MHz, $a \ll 100$ MHz).

An interesting intermediate case is that of ^{113}Cd, where a and Γ are of the same order. Lehmann has shown that in this case the study of the efficiency of the optical pumping process permits measuring a in spite of the fact that the HFS remains unresolved.

Note that in the case of ^{67}Zn the optical resonance light emitted is completely σ^+ polarized. There is no π light present. (The polarization of the resonance radiation is the same as if there were no nuclear spin.) The measurement of the ratio of intensities π/σ in the emitted resonance radiation gives information on a/Γ.

Other Tests of the Franck-Condon Principle

We have stated that the nuclear-spin axis remains unchanged not only in spectral transitions but also during sudden collision processes which cause a perturbation of the electron configuration. Again the study of polarization of resonance radiation of ^{199}Hg gives us a method of testing this prediction.

The following experiment has been made by Faroux: By irradiating ^{199}Hg vapor with σ^--polarized light of ^{204}Hg, he excites selectively the ^{199}Hg atoms to level $m_F = -\frac{1}{2}$ of the hyperfine state $F = \frac{1}{2}$ of 6^3P_1 according to the scheme of Figure 4-4. Faroux adds to the ^{199}Hg vapor in the resonance cell a small quantity of He gas, just enough so that,

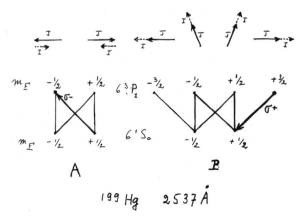

Fig. 4-4. Collisions in the excited state $6\,^3P_1$ of the ^{199}Hg atom with He atoms; transfer from hyperfine state $F = \frac{1}{2}$ to $F = \frac{3}{2}$; for ^{199}Hg, 2537 Å. The nuclear orientation (i vector) remains unchanged.

during the life time of the excited Hg atoms in state 6^3P_1, a small proportion of these atoms is submitted to collisions with He atoms.

Such collisions produce transfers to other Zeeman levels and also transfers from the hyperfine level $F = \frac{1}{2}$ to the level $F = \frac{3}{2}$. (Hyperfine component B, $F = \frac{3}{2} \rightarrow i = \frac{1}{2}$, appears in the emitted resonance radiation as was established by Mrozowski many years ago.) It is well known that foreign-gas collisions produce depolarization of the resonance radiation. In Faroux' experiment the He pressure is so small that the chance for an atom in the excited state to collide a second time with a He atom is negligeably small. We apply the Franck-Condon principle to the one-collision process. Above the Zeeman scheme of the a and b components we have indicated, according to the old quantum theory, the orientation of the F vector ($\mathbf{F} = \mathbf{J} + \mathbf{i}$) for the six m_F levels of the upper state. On this vector model, we see that the conservation of orientation of the i vector in a collision excludes a transfer from level $F = \frac{1}{2}, m_F = -\frac{1}{2}$ to $F = \frac{3}{2}, m_F = -\frac{3}{2}$ but that it permits transfers to level $F = \frac{3}{2}, m_F = +\frac{3}{2}$. We conclude that in the emission of the B component the intensity of σ^+ light must be dominant. The collision transfer $F = \frac{1}{2} \rightarrow F = \frac{3}{2}$ is associated with the reversal of circular polarization of the σ light.

Figure 4-5 shows the quantitative results of a quantum mechanical calculation made by Faroux for the one-collision transfer. The above conclusion, drawn from the vector model, is confirmed. If state $F = \frac{1}{2}$ is selectively excited by σ^- light, the polarization ratio of the B component produced in the one-collision transfer process should be $\mathscr{T}_{\sigma^+}/\mathscr{T}_{\sigma^-} = \frac{23}{13}$.

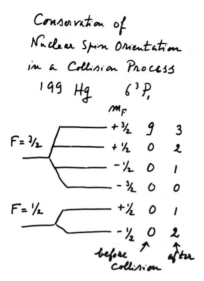

Fig. 4-5. Conservation of nuclear spin orientation in a collision process in [199]Hg, 6 [3]P$_1$, showing populations before and after collision (Faroux and Brossel 1966).

This prediction has been experimentally confirmed by Faroux. In the scattered resonance light the A component can be eliminated by a [204]Hg filter, and the polarization of the isolated B component can be measured. It is partially circularly polarized in a sense opposite to the exciting A light. Figure 4-6 shows the result. Faroux measured the Faraday rotation produced by the vapor of the resonance cell near the A line and near the B line. He observed reversed signs for the Faraday rotation.

Use of the Franck–Condon Principle

We may describe nuclear orientation by the optical pumping process in the following way: We start with an atomic ensemble which is isotropic in the electron configuration and in the nuclear situation:

$$J = 0 \qquad \langle J_z \rangle = 0 \qquad \langle I_z \rangle = 0$$

By spectral transitions, we introduce electronic dissymmetry:

$$J \neq 0 \qquad \langle J_z \rangle \neq 0 \qquad \langle I_z \rangle = 0$$

The hyperfine coupling $a\mathbf{JI}$ produces a transfer of orientation from J to I:

$$J \neq 0 \qquad \langle J_z \rangle \neq 0 \; . \quad \langle I_z \rangle \neq 0$$

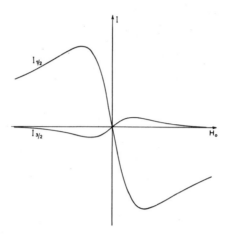

Fig. 4-6. Collisions in the excited state $6\ ^3P_1$ of ^{199}Hg with He; transfer from hyperfine state $F = \frac{1}{2}$ to $F = \frac{3}{2}$; inversion of the sign of m_F (circular polarization) by the transfer (Faroux and Brossel 1966).

A last spectral transition (emission) suppresses the J orientation. The nuclear orientation remains:

$$J = 0 \qquad \langle J_z \rangle = 0 \qquad \langle I_z \rangle \neq 0$$

The transfer of orientation through the hyperfine coupling can be used in the opposite way: We start from a situation where we have achieved nuclear orientation, for example, in a ground state:

$$J = 0 \qquad \langle J_z \rangle = 0 \qquad \langle I_z \rangle \neq 0$$

By isotropic electronic excitation (an isotropic gas discharge), we create a situation where

$$J \neq 0 \qquad \langle J_z \rangle = 0 \qquad \langle I_z \rangle \neq 0$$

The hyperfine coupling produces a transfer of orientation from $\langle I \rangle$ to $\langle J \rangle$:

$$J \neq 0 \qquad \langle J_z \rangle \neq 0 \qquad \langle I_z \rangle \neq 0$$

The spectral lines originating from this state $\langle J_z \rangle \neq 0$ will give rise to polarized light emission, and this may be used to monitor magnetic resonance in excited states of atoms (Fig. 4-7).

This method has been applied by Laloë to the study of magnetic resonance in excited states of ^{3}He: nuclear orientation is achieved in the ground state of ^{3}He by the process of metastability exchange (10% orientation can be easily obtained after 10 minutes of operation). The

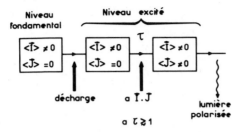

Fig. 4-7. Principle of transfer of nuclear orientation to electronic orientation in the excitation process of an atom (Laloë 1968).

He atoms are reexcited in an isotropic gas discharge, and the spectral lines emitted appear polarized.

Figure 4-8 shows some of the lines studied in this way. The circular polarization of line 6678 Å originating from level 3^1D is particularly strong. Figure 4-9 shows, on this line, the build-up of polarization during the nuclear orientation process. In a zero magnetic field the $\langle J_z \rangle$ assymmetry produced in this way is pure orientation; there is no alignment. Application of a magnetic field produces alignment as is shown on Figure 4-10. The spectral lines then show linear polarization when observed in a direction perpendicular to the field and to the orientation axis. Figure 4-11 shows the destruction of orientation or of alignment by magnetic resonance (optical signals on $\lambda = 6678$ Å), and Table 4-1 shows some results obtained by Laloë: measurement of the HFS of three ^1D states of ^{3}He.

Table 4-1. Hyperfine structures (MHz) of the ^{3}He atom according to Pavlović and Laloë (1969) where the references to Moser's and Descoubes' works are cited

State	Moser's Calculations [41]	Descoubes' Measurements [12, 30]	Measurements by Pavlović and Laloë
3^1D	146 ± 3	86.8 ± 1.5 109 ± 2 *138 ± 2.5	136 ± 3
4^1D	110 ± 3	64.3 ± 0.6 80.8 ± 0.8 *102 ± 1	100 ± 3
5^1D	103 ± 5	59.0 ± 0.6 74.2 ± 0.7 *93.8 ± 0.9	94 ± 3

* According to the first column these values have to be selected.

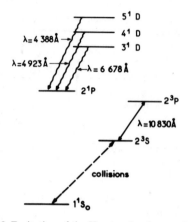

Fig. 4-8. Excitation of the He atom by electron impact.

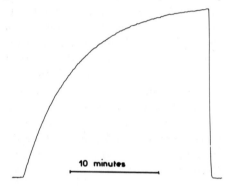

Fig. 4-9. Build-up of the orientation process of ^{3}He as a function of time (Laloë 1968).

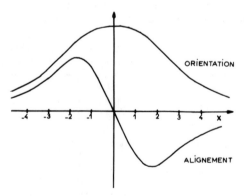

Fig. 4-10. Orientation and alignment produced in the excited state $3\,^1D_2$ of ^{3}He (Pavlović and Laloë 1969).

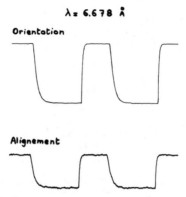

λ = 6.678 Å

Orientation

Alignement

Fig. 4-11. Orientation and alignment signals on $\lambda = 6678$ Å originating from the level 3 ^{1}D of ^{3}He (Laloë 1968).

PRODUCTION, MEASUREMENT, AND USE OF VERY SMALL MAGNETIC FIELDS

In Chapter 2, when we studied the Hanle effect, we saw that the half-width of the Hanle curve in the magnetic field scale is given by:

$$H_c = \gamma^{-1}\tau^{-1}$$

where γ is the gyromagnetic ratio of the atomic state considered and τ its decay time. If we apply this formula to an alkali vapor in a paraffin-coated cell oriented by optical pumping, the numerical values are:

$$\gamma = 10^7 \qquad \text{and} \qquad \tau \approx 1 \text{ sec}$$

which gives $H_c \approx 10^{-7}$ gauss.

Such a critical field, which reduces the ordinate of the Hanle curve to half of its zero-field maximum, is much smaller than the stochastic fluctuations of the earth field in a laboratory. So the first condition to satisfy to study the Hanle curve of optically oriented alkali atoms is to reduce the earth field and its erratic fluctuations. This has been achieved by Haroche using five concentric screening cylinders of μ metal. Figure 4-12 shows Haroche's experimental arrangement for ^{87}Rb, and Figure 4-13 shows a Hanle curve obtained by him. The small field H_0 is produced by a small calibrated current in a Helmholtz coil placed inside the screens.

The scale shows that the half-width of the curve is of the order of 1μG = 1 microgauss = 10^{-6} gauss, larger than expected. This is because the pumping light shortens the relaxation time of the ground state. A

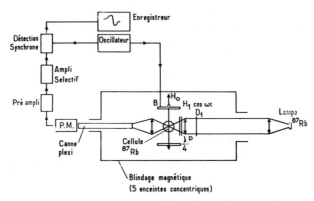

Fig. 4-12. Experimental arrangement for registration of the Hanle curve of the ground state of ^{87}Rb (Cohen-Tannoudji, Dupont-Roc, and Haroche 1969).

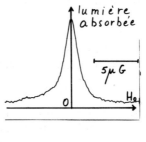

Fig. 4-13. Hanle effect in the ground state of ^{87}Rb (Cohen-Tannoudji, Dupont-Roc, and Haroche 1969).

better signal-to-noise ratio can be obtained by using the a-c modulation of the Hanle effect described in Chapter 2, which gives the dispersion curve shown in Figure 4-14. Figure 4-15 shows that the signal-to-noise ratio is excellent and that the curve near zero field is very steep. The steepness can be used to detect very small changes of the H_0 field in the neighborhood of the zero field (Fig. 4-16).

Figure 4-17 shows signals obtained by a change of $2 \cdot 10^{-9}$ gauss. With a multichannel technique still smaller field changes can be monitored (Fig. 4-18).

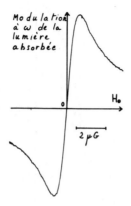

Fig. 4-14. Modulation signal of the Hanle curve of ^{87}Rb with small light intensity. A magnetic RF field $H_1 \cos \omega t$ parallel to the steady field H_0 is applied. The modulation is obtained on frequency ω (Cohen-Tannoudji, Dupont-Roc, and Haroche 1969).

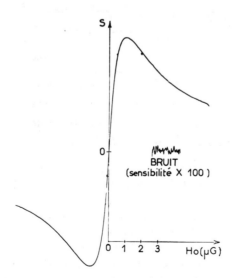

Fig. 4-15. Modulation signal of the Hanle curve of ^{87}Rb with optimum light intensity to obtain the best signal-to-noise ratio (Cohen-Tannoudji, Dupont-Roc, and Haroche 1969).

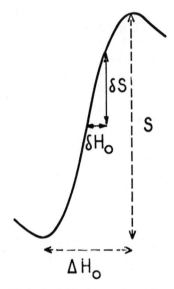

Fig. 4-16. A small change δH_0 in the field value produces a large change δS in the light signal (Cohen-Tannoudji, Dupont-Roc, and Haroche 1969).

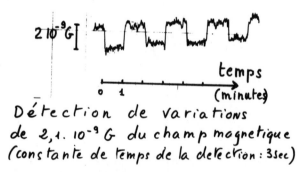

Fig. 4-17. Response of the signal to a square pulse of magnetic field of 2.1×10^{-9} gauss with time constant 3 sec (Dupont-Roc, Haroche, and Cohen-Tannoudji 1969).

By Haroche's method very small fields can be produced and measured. What can we do with such small fields?

Let me give you an example of application to atomic physics. Near the rubidium cell which detects such small fields, Haroche placed a cell of diameter 6 cm filled with ^{3}He gas at a pressure of 3 Torr (Figs. 4-19 and 4-20). By optical pumping and metastability exchange 5% of the nuclei

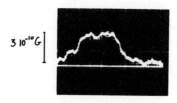

Reponse du signal à des pulses
carrés périodiques de champ magnétique
d'amplitude 3.10⁻¹⁰ G.
    Détection par "analyseur multicanaux"
(3000 passages; constante de temps de la
détection: 0,1 s)

Fig. 4-18. Response of the signal to repetitive square pulses of magnetic field of 3×10^{-10} gauss in amplitude; 3,000 runs, time constant 0.1 sec (Dupont-Roc, Haroche, and Cohen-Tannoudji 1969).

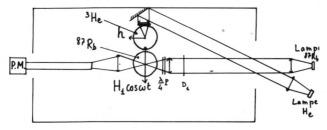

Fig. 4-19. Detection of the static magnetic field produced by the oriented nuclei of optically pumped ^{3}He gas. Experimental arrangement by Haroche.

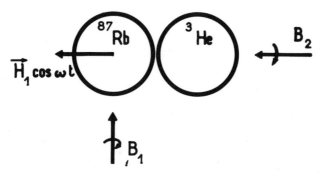

Fig. 4-20. Detail of the experimental arrangement of Figure 4-19 by Haroche.

of the ^{3}He atoms can be oriented in a direction pointing to the center of the Rb cell. A calculation shows that the field produced by this nuclear magnetization at the center of the Rb cell is of the order of $6 \cdot 10^{-8}$ gauss, largely strong enough to be detected. Figure 4-21 shows the build-up of this magnetization monitored by this field. If now a field H is applied

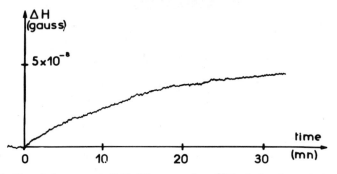

Fig. 4-21. Plot of the magnetic field ΔH measured on ^{87}Rb during the optical pumping process of ^{3}He, with ^{3}He pressure of 3 Torr (Cohen-Tannoudji, Dupont-Roc, Haroche, and Laloë 1969).

to the He cell in a direction perpendicular to the magnetization M, this magnetization will precess with the Larmor frequency around field H, and this precession will modulate the field seen by the Rb atoms. In a field $H = 2 \cdot 10^{-6}$ gauss the precession frequency is $\nu = 6 \cdot 10^{-3}$ Hz,

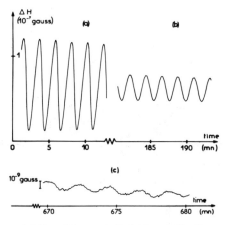

Fig. 4-22. Plot of the modulation of the magnetic field caused by the precession of the nuclear spins of ^{3}He: (a) just after optical pumping has been stopped; (b) 3 hr later; 3 Torr; $P = 5\%$; (c) 11 hr later, enhanced sensitivity (Cohen-Tannoudji, Dupont-Roc, Haroche, and Laloë 1969).

which corresponds to a period of precession of about 2 minutes. Figure 4-22 shows the modulation produced by the precession of the ^{3}He nuclei, monitored by the Rb atoms. This experiment enables us to measure the nuclear relaxation time T_1 of ^{3}He in a glass bulb. The measurement gives $T_1 = 2$ hours 20 minutes.

REFERENCES

Cohen-Tannoudji, C.; Dupont-Roc, R.; and Haroche, S. 1969. Colloquium on Weak Magnetic Fields, C.N.E.S. Paris, May 1969. To be published in *J. de Physique* supplement.

Cohen-Tannoudji, C.; Dupont-Roc, J.; Haroche, S.; and Laloë, F. 1969. *Phys. Rev. Letters* 22:758.

Dupont-Roc, J.; Haroche, S.; and Cohen-Tannoudji, C. 1969. *Phys. Letters* 28A:638.

Faroux, J. P., and Brossel, J. 1966. *Compt. Rend.* Acad. Sci. Paris 262:41.

Laloë, F. 1968. *Compt. Rend.* Acad. Sci. Paris 267:208.

Pavlović, M., and Laloë, F. 1969. *Compt. Rend.* Acad. Sci. Paris 268:1436, 1485.

5

ADNAN ŞAPLAKOĞLU

Atomic and Molecular Beam Experiments

I shall summarize the main features of the atomic beam magnetic resonance method to provide an introduction for the subsequent chapter on the subject. Furthermore, at least for nonatomic beamists, a look into what and how things are measured experimentally will form a bond between the areas of calculation and measurement.

I shall also discuss as an example the atomic beam machine of the Orta-Doğu Teknik Üniversitesi in Turkey, which has been constructed in our Physics Department.

APPARATUS

Experimental studies made with atomic and molecular beams enable us to measure the various atomic and molecular constants to a high degree of accuracy. Electronic and nuclear Lande g factors g_J, g_I; electronic and nuclear magnetic moments μ_J, μ_I; magnetic dipole, electric quadrupole, and magnetic octupole interaction constants a, b, and c; hyperfine structure separation Δv; and hyperfine structure anomaly $_1\Delta_2$ are such constants. All of the above are quite relevant to a test of the theory.

Beam studies were first started by Dunoyer (1911a, b). His purpose was to deposit Na atoms on a surface to form a thin film.

Beam technique was further developed by Stern and collaborators in Hamburg between 1920 and early 1929.

At the beginning, beams were used only as a method to measure nuclear magnetic moments. In 1938 the molecular beam resonance method was introduced by Rabi, Zacharias, Millman, and Kusch (1938) and by Kellogg, Rabi, Ramsey, and Zacharias (1939a, b).

Afterward the more general technique of radiofrequency spectroscopy was developed for molecules by Kellogg, Rabi, Ramsey, and Zacharias (1939a, b) and later this was used for the study of atomic states (Kusch, Millman, and Rabi 1940).

Essential features of this technique have been used since then. The idea was to introduce a radiofrequency oscillating magnetic field while the beam was inside a homogeneous magnetic field. Before going into how this is done, we shall look very briefly into the formation of the beam.

The beam is produced in a heated oven, which is a small cup containing a narrow slit in front. The sample to be studied is put into the oven in small quantities and the oven is heated by electron bombardment or by simple heating wires.

The velocity distribution inside the oven is Maxwellian and given by

$$\frac{dN}{dv} = \frac{4N}{\sqrt{\pi}} \frac{1}{\alpha^3} \cdot v^2 \exp\left(-v^2/\alpha^2\right)$$

where N is the total number of molecules in the oven. Since the effusion of molecules from the oven slit is proportional to the velocity, the beam intensity becomes

$$I(v) = \frac{2I_0}{\alpha^4} \cdot v^3 \exp\left(-v^2/\alpha^2\right)$$

where $\alpha = \sqrt{2kT/m}$ is the most probable molecular velocity.

Dealing with the free atoms or molecules in beam form like this has some advantages. In the first place there is no effect of surrounding atoms since individual atoms in the beam are far enough from each other. Secondly, outside interactions to be imposed on the atom do not have the inconvenience of having to penetrate through a container, since no container exists.

Enough about the beam. Let us bring it into the main part of the atomic beam machine. As you see from Figure 5-1 (top view), the beam is allowed to cross a homogeneous magnetic field H perpendicular to its direction of motion. On its way, it passes through a single loop coil and arrives at the detector.

The outside interaction in our case is produced by this small loop, called a hairpin, which carries a radiofrequency current of frequency v.

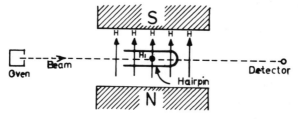

Fig. 5-1. Top view of the homogeneous magnetic field region of the atomic and molecular beam machine.

This hairpin produces an oscillating magnetic field H_1 perpendicular both to the external magnetic field H and to the direction of the beam. (Perpendicular to the paper and shown at an instant when heading upward.)

It can be shown that the oscillating magnetic field H_1 is equivalent to two rotating magnetic fields of magnitude $H_1/2$, each rotating in opposite directions in a plane perpendicular to H.

To better visualize a classical approach, for a simple case, an atom having only a nuclear magnetic moment μ_I will be discussed.

The nucleus of magnetic moment μ_I will be precessing about H with Larmor angular frequency

$$\omega_0 = \frac{\mu_I}{\hbar I} H$$

and at the same time it is under the influence of the small magnetic field $H_1/2$ which is rotating in the correct sense, where I is nuclear spin. If the angular frequency of the rotating magnetic field is equal to the Larmor precessional angular frequency ω_0, it is obvious that the plane produced by μ_I and $H_1/2$ will maintain its orientation along the magnetic field H.

This continuous torque will then cause a shift in the direction of the magnetic moment into one of the next allowed values of orientation.

By measuring this transition frequency and using the known value of H it becomes possible to measure the energy separation corresponding to the two different allowed orientations of the magnetic moment in a magnetic field. The simple Bohr condition will then lead us to the value of μ_I.

THEORY

If we go step by step, looking into the main part of the machine in Figure 5-1, we shall first look at the free atoms, then we shall impose a magnetic field H on them, and finally we shall apply an oscillating magnetic field H_1.

Free Atom Hamiltonian

Electrostatic Interaction. The electrostatic interaction between the electrons and nuclei can be calculated from the interaction between the electronic and nuclear volume elements $d\tau_e$ and $d\tau_n$ separated by a distance r, as shown in Figure 5-2.

If the electronic charge density at the position of $d\tau_e$ is ρ_e and the nuclear charge density at the position of $d\tau_n$ is ρ_n, this interaction is

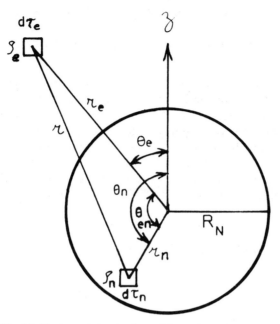

Fig. 5-2. Electrostatic interactions of the nucleus with electrons.

given by $(\rho_e \rho_n \, d\tau_e \, d\tau_n)/r$. Upon integrating over the whole space, the electrostatic interaction is obtained as

$$\mathscr{H}_E = \int\limits_{\tau_e} \int\limits_{\tau_n} \frac{\rho_e \rho_n \, d\tau_e \, d\tau_n}{r}$$

Replacing $1/r$ in terms of electronic and nuclear distances from the centroid of the nucleus from Figure 5-2, we get

$$\frac{1}{r} = (r_e^2 + r_n^2 - 2r_e r_n \cos \theta_{en})^{-\frac{1}{2}}$$

$$= \frac{1}{r_e} P_0 + \frac{r_n}{r_e^2} P_1 + \frac{r_n^2}{r_e^3} P_2 + \cdots$$

where

$$P_k = \frac{1}{2^k k!} \frac{d^k}{d(\cos \theta_{en})^k} (\cos^2 \theta_{en} - 1)^k$$

the kth degree Legendre polynomial. Substituting, we get

$$\mathscr{H}_E = \sum_k \mathscr{H}_{Ek} = \sum_k \int_{\tau_e} \int_{\tau_n} \frac{\rho_e^e \rho_n}{r_e} \left(\frac{r_n}{r_e}\right)^k P_k(\cos\theta_{en})\, d\tau_e\, d\tau_n$$

where ρ_e^e means the electronic charges outside the nuclear sphere of radius R_N: This is not the case for an s electron, and the wavefunction does not vanish at the nucleus. For this reason I would like to note that this bypass will be partly responsible for the hyperfine structure anomaly according to the Bohr–Weisskopf theory (Bohr and Weisskopf 1950; Bohr 1951). The disadvantage of this representation is the θ_{en} dependence rather than dependence of θ_e and θ_n separately. If we use the spherical harmonic addition theorem, this difficulty can be resolved and $\mathscr{H}_{Ek}$ is expressed by

$$\mathscr{H}_{Ek} = \mathbf{Q}^{(k)} \cdot \mathbf{F}^{(k)}$$

where $\mathbf{Q}^{(k)}$ and $\mathbf{F}^{(k)}$ are irreducible tensors of degree k and the dot in between means

$$\mathscr{H}_{Ek} = \sum_{q=-k}^{k} (-1)^q Q_q^{(k)} F_{-q}^{(k)}$$

where $Q_q^{(k)}$ and $F_{-q}^{(k)}$ are given by,

$$Q_q^{(k)} = \int_{\tau_n} \rho_n r_n^k C_q^{(k)}(\theta_n, \phi_n)\, d\tau_n$$

$$F_q^{(k)} = \int_{\tau_e} \rho_e^e r_e^{-(k+1)} C_q^{(k)}(\theta_e, \phi_e)\, d\tau_e$$

In the above equations ϕ_n and ϕ_e are azimuthal angles from the xy plane, and $C_q^{(k)}$, with a normalization factor difference, is equal to the normalized tesseral harmonic

$$C_q^{(k)}(\theta, \phi) = \left(\frac{4\pi}{2k+1}\right)^{\frac{1}{2}} Y_q^{(k)}(\cos\theta, \phi)$$

Then the nuclear multipole electric interaction becomes (Table 5-1):

$$\mathscr{H}_E = \mathscr{H}_{E0} + \mathscr{H}_{E1} + \mathscr{H}_{E2} + \mathscr{H}_{E3} + \mathscr{H}_{E4} + \cdots$$

From parity considerations, however, it can be shown that no odd electrical multipole moment can exist. If odd k terms are crossed out

$$\mathscr{H}_E = \mathscr{H}_{E0} + \mathscr{H}_{E2} + \mathscr{H}_{E4} + \cdots$$

Table 5-1. Nuclear multipole electric interaction

Monopole	Dipole	Quadrupole	Octupole	16-pole
$k = 0$	$k = 1$	$k = 2$	$k = 3$	$k = 4$
$2^0 = 1$	$2^1 = 2$	$2^2 = 4$	$2^3 = 8$	$2^4 = 16$

The electrical dipole moment has been reinvestigated recently at Brookhaven National Laboratory (Cohen, Nathans, Silsbee, Lipworth, and Ramsey 1968). The result showed that, if a neutron has an electrical dipole moment, the separation between the $+e$ and $-e$ charges must be smaller than $(-2 + 5)10^{-22}$ cm. This confirms the theoretical predictions.

There are also theoretical restrictions on the order of electric multipole moments: If the nuclear spin is I, the multipole moment of the order 2^k, $k > 2I$ cannot exist. That is, for example, for a nucleus of spin $\frac{3}{2}$, $2I = 3$, and the highest order for the electric moment is $2^3 = 8 = $ octupole moment. But, since $\mathscr{H}_{E3}$ is eliminated by the first theorem, the highest possible moment is the electric quadrupole moment.

For spin $I = \frac{1}{2}$, no electric moment is possible other than the zero order monopole moment, which is nothing but the Coulomb interaction.

Magnetic Interaction. Let us indicate the nuclear current density by $\mathscr{I}_n$, and the electron current density by $\mathscr{I}_e$.

Then $\mathscr{I}_e$ will produce a vector potential $\mathbf{A}_e$ at the nucleus and $\mathscr{I}_n$ will produce a vector potential $\mathbf{A}_n$. The mutual potential energy for the magnetic interaction between the nucleus and the electrons becomes

$$\mathscr{H}_M = -\frac{1}{c} \int_{\tau_n} \mathscr{I}_n \mathbf{A}_e \, d\tau_n = -\frac{1}{c} \int_{\tau_e} \mathscr{I}_e \mathbf{A}_n \, d\tau_e$$

Using the continuity equation

$$\nabla \cdot \mathscr{I} = -\dot{\rho} = 0$$

and deriving $\mathscr{I}_n$ from a vector potential

$$\mathscr{I}_n = c\nabla \times \mathbf{m}_n$$

leads to the result

$$\mathscr{H}_M^e = \int_{\tau_e} \int_{\tau_n} \frac{(-\nabla_n \mathbf{m}_n)(-\nabla_e \mathbf{m}_e)}{r} \, d\tau_e \, d\tau_n$$

This is the same as $\mathcal{H}_E$ except for the replacements,

$$\rho_n \to -\mathbf{V}_n\mathbf{m}_n \qquad \text{and} \qquad \rho_e \to -\mathbf{V}_e\mathbf{m}_e$$

The results obtained for the electrostatic interaction are applicable with the above correspondence. The magnetic interaction Hamiltonian becomes:

$$\mathcal{H}_M = \mathcal{H}_{M0} + \mathcal{H}_{M1} + \mathcal{H}_{M2} + \mathcal{H}_{M3} + \cdots$$

Similar restrictions on the multipole magnetic moments are that no even multipole magnetic moment can exist, and no multipole magnetic moment of order 2^k can exist for $k > 2I$. That is, $k \leqslant 2I$ always.

Then the magnetic interaction becomes

$$\mathcal{H}_M = \mathcal{H}_{M1} + \mathcal{H}_{M3} + \cdots$$

and the total interaction Hamiltonian is

$$\mathcal{H} = \mathcal{H}_E + \mathcal{H}_M = \mathcal{H}_{E0} + \mathcal{H}_{M1} + \mathcal{H}_{E2} + \mathcal{H}_{M3} + \cdots$$

Since in RF spectroscopy we deal with the differences between the energy states, we do not need the orientation independent terms in the Hamiltonian.

Let us calculate $\mathcal{H}_{M1}$, the magnetic dipole interaction Hamiltonian. That is, the interaction between the magnetic dipole moment of the nucleus with the magnetic field due to the angular momentum of electrons.

The angular momentum of a nucleus is indicated by **a** in units of erg. sec. In order to deal with unitless quantities we separate $\hbar$ from it and call the other factor **I**. We then have $\mathbf{a} = \hbar\mathbf{I}$, so that **I** measures the angular momentum in units of $\hbar$. The maximum possible component of **I** in any given direction is I and is called the spin of the nucleus. The magnetic moment vector $\boldsymbol{\mu}_I$ is defined as proportional to the angular momentum **a**, so that

$$\boldsymbol{\mu}_I = \gamma_I\hbar\mathbf{I} = g_I\mu_N\mathbf{I}$$

where γ_I is the gyromagnetic ratio and g_I is the Lande g factor. The maximum possible component of $\boldsymbol{\mu}_I$ is called the magnetic moment μ_I and is a scalar quantity. Then

$$\mu_I = \gamma_I\hbar I \qquad \text{and} \qquad \boldsymbol{\mu}_I = \frac{\mu_I}{I}\mathbf{I}$$

The magnetic dipole interaction becomes

$$\mathcal{H}_{M1} = -\boldsymbol{\mu}_I \cdot \mathbf{B}_J$$

where $\mathbf{B}_J$ is the magnetic induction produced at the nucleus by the circulating and spinning electrons. Corrections for the variations of $\mathbf{B}_J$ due to the nuclear magnetization over the volume of the nucleus are still needed. This is rather small, at the worst (heavy nuclei and s electrons) a few percent. However, this is responsible for the hyperfine-structure anomaly. Replacing the value of $\boldsymbol{\mu}_I$

$$\mathscr{H}_{M1} = -\frac{\mu_I}{I}\mathbf{I} \cdot \mathbf{B}_J$$

Similar to the case for the $\boldsymbol{\mu}_I$ and $\mathbf{I}$ proportionality we can express $\mathbf{B}_J$ as proportional to $\mathbf{J}$, for matrix elements diagonal in J, and we obtain

$$\mathscr{H}_{M1} = ha\mathbf{I} \cdot \mathbf{J}$$

where

$$ha = -\frac{\mu_I}{I}\frac{\mathbf{B}_J \cdot \mathbf{J}}{\mathbf{J} \cdot \mathbf{J}}$$

and where a is called the hyperfine-structure, magnetic-dipole interaction constant. When the interaction between $\mathbf{I}$ and $\mathbf{J}$ is large in comparison with the interactions between the external field, the resultant angular momentum quantum number F is a good quantum number, and the energy W_{M1} for the hyperfine structure $\mathscr{H}_{M1} = ha\mathbf{I} \cdot \mathbf{J}$ can be calculated in the Fm representation by the aid of Figure 5-3.

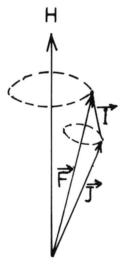

Fig. 5-3. Coupling of electronic and nuclear angular momentum vectors.

$$\mathbf{F}^2 = \mathbf{J}^2 + \mathbf{I}^2 + 2\mathbf{I} \cdot \mathbf{J}$$

$$\mathbf{I} \cdot \mathbf{J} = \tfrac{1}{2}[F(F + 1) - I(I + 1) - J(J + 1)]$$

$$W_{M1} = \frac{ha}{2}[F(F + 1) - I(I + 1) - J(J + 1)]$$

If the energy difference

$$\Delta W = W_{M1}(F) - W_{M1}(F - 1) = haF$$

then the separation in frequency becomes

$$\Delta v = \frac{|\Delta W|}{h} = aF$$

where Δv is known as the hyperfine structure separation. For example for $J = \tfrac{1}{2}$ and arbitrary I

$$\Delta v = a(I + \tfrac{1}{2})$$

For the hydrogenlike atoms the effective magnetic induction $\mathbf{B}_J$ at the nucleus is calculated quantum mechanically (Fermi 1930; Goudsmit 1929, 1931; Kopfermann 1940). We shall give a semiclassical treatment below. However we would like to note that the calculation differs more for s electrons than for any other electrons because the s electron wavefunction does not vanish at the nucleus. The magnetic induction $\mathbf{B}_J$ at the nucleus can be written as

$$\mathbf{B}_J = \mathbf{B}_{J1} + \mathbf{B}_{J2}$$

where $\mathbf{B}_{J1}$ is the magnetic induction caused by the electron density inside the nucleus and $\mathbf{B}_{J2}$ is the magnetic induction caused by the electron density outside the nucleus.

For an S state $\mathbf{B}_{J2} = 0$ because of spherical symmetry. If $\mathbf{m}_e$ is the magnetization of the nucleus, $\mathbf{B}_J$ is given by

$$\mathbf{B}_J = \mathbf{B}_{J1} = \frac{8\pi}{3}\mathbf{m}_e$$

where

$$\mathbf{m}_e = -|\psi_{n0}(0)|^2 2\mu_0 S = -|\psi_{n0}(0)|^2 2\mu_0 \cdot \mathbf{J}$$

where $\psi_{n0}(0)$ is the wavefunction of the s electron at the centroid of the nucleus and μ_0 is the Bohr magneton. Therefore

$$\mathbf{B}_J = -\frac{16\pi}{3}\mu_0|\psi_{n0}(0)|^2\mathbf{J}$$

and

$$ha = \frac{16\pi}{3}\mu_0\frac{\mu_I}{I}|\psi_{no}(0)|^2 = \frac{8}{3}\frac{hc\,\text{Ry}\,\alpha^2 Z^2 g_I}{n^3(M/m)}$$

where $\alpha = e^2/\hbar c$ is the fine structure constant and Ry is the Rydberg constant. For non s states ha is given by Ramsey (1953):

$$ha = \frac{hc\,\text{Ry}\,\alpha^2 Z^3 g_I}{n^3(M/m)(L + \frac{1}{2})(J + 1)J}$$

The electric quadrupole interaction is given by (Ramsey 1953),

$$\mathscr{H}_{E2} = hb\frac{3(\mathbf{I}\cdot\mathbf{J})^2 + \frac{3}{2}\mathbf{I}\cdot\mathbf{J} - \mathbf{I}^2\mathbf{J}^2}{2I(2I-1)J(2J-1)}$$

where b is the hyperfine-structure, electric-quadrupole interaction constant and is related to the nuclear quadrupole moment Q by

$$hb = e^2 q_J Q$$

where q_J is given by

$$q_J = \frac{1}{e}\left\langle\frac{\partial^2 v^e}{\partial z^2}\right\rangle_{JJ}$$

The angle brackets indicate the average potential and the subscript JJ means that the molecule is in the state of $m_J = J$. Our total Hamiltonian becomes

$$\mathscr{H} = ha\mathbf{I}\cdot\mathbf{J} + hb\frac{3(\mathbf{I}\cdot\mathbf{J}) + \frac{3}{2}\mathbf{I}\cdot\mathbf{J} - I(I+1)J(J+1)}{2I(2I-1)J(2J-1)}$$

If we introduce the external field interactions, $\mathscr{H}$ becomes

$$\mathscr{H} = ha\mathbf{I}\cdot\mathbf{J} + hb\frac{3(\mathbf{I}\cdot\mathbf{J}) + \frac{3}{2}\mathbf{I}\cdot\mathbf{J} - I(I+1)J(J+1)}{2I(2I-1)J(2J-1)}$$

$$- g_J\mu_0\mathbf{J}\cdot\mathbf{H} - g_I\mu_0\mathbf{I}\cdot\mathbf{H}$$

Obtaining the energy levels as a function of H is rather straightforward and in low H fields the Fm representation is adequate because $\mathbf{L}$ and $\mathbf{S}$ couple to give $\mathbf{J}$, and $\mathbf{J}$ and $\mathbf{I}$ couple to give $\mathbf{F}$. If we ignore the electric quadrupole term, the matrix elements for

$$W_{M1}(F, m) = \langle Fm|\mathscr{H}_{M1}|Fm\rangle$$

can be readily obtained. At high fields $\mathbf{J}$ and $\mathbf{I}$ decouple as shown in Figure 5-4 and precess around H independently. In that case the ms are

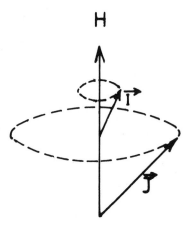

Fig. 5-4. Decoupling of the electronic and nuclear angular momentum vectors at high magnetic field.

not good quantum numbers anymore, and the $m_I m_J$ representation is adequate. The matrix elements

$$W_{M1}(m_I, m_J) = \langle m_I m_J | \mathscr{H}_{M1} | m_I m_J \rangle$$

can be obtained easily.

For intermediate magnetic fields, however, the secular equation must be solved. The Hamiltonian of the system is given by

$$\mathscr{H} = ha I_z J_z + \tfrac{1}{2} ha I_+ J_- + \tfrac{1}{2} ha I_- J_+ - \frac{\mu_J}{J} J_z H - \frac{\mu_I}{I} I_z H$$

where $I_\mp = I_x \mp i I_y$ and $J_\mp = J_x \mp i J_y$. The general solution of the secular equation for arbitrary I and J is quite lengthy, but a solution for $J = \tfrac{1}{2}$ and arbitrary I is quite simple and illuminating. The result is given by

$$W_{M1}(F, m) = -\frac{h\Delta v}{2(2I + 1)} - \frac{\mu_I}{I} H m \mp \frac{h\Delta v}{2} \left(1 + \frac{4m}{2I + 1} x + x^2 \right)^{\tfrac{1}{2}}$$

where

$$h\Delta v = haF = ha(I + \tfrac{1}{2}) \qquad \text{and} \qquad x = \frac{(-\mu_J/J + \mu_I/I)H}{h\Delta v}$$

where the plus sign for $F = I + \tfrac{1}{2}$ and the minus sign for $F = I - \tfrac{1}{2}$ should be used. A plot of this equation for $I = \tfrac{3}{2}$ is shown in Figure 5-5, which is known as the Breit–Rabi diagram.

Now that we have obtained the energy level diagram as a function of x, a quantity related to the homogeneous magnetic field H, we are ready to perform the experiment. The selection rules for the allowed transitions are:

$$\Delta m_I = 0, \mp 1; \quad \Delta m_J = 0, \mp 1 \quad \text{in strong external fields}$$

$$\Delta F = 0, \mp 1; \quad \Delta m = 0, \mp 1 \quad \text{in weak external fields}$$

Some of the allowed transitions between the energy levels corresponding to an arbitrary magnetic field are shown with vertical dashed arrows of Figure 5-5.

Transitions with $\Delta m = 0$ occur when there is a component of the oscillatory field H_1 in the direction of the external field H. Transitions of this kind are called σ transitions.

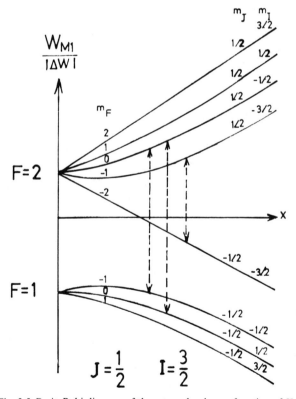

Fig. 5-5. Breit–Rabi diagram of the energy levels as a function of H.

The transitions in which m changes by ∓ 1 are called π transitions and are induced by the oscillatory field H_1 perpendicular to the direction of H.

EXPERIMENTAL PROCEDURE

A schematic diagram of the atomic-beam machine is shown in Figure 5-6. Transitions are induced by the RF oscillating field in the homogeneous H field region called the C field.

This whole system, including the oven and detector, is held in a vacuum of about $5 \cdot 10^{-6}$ mm Hg.

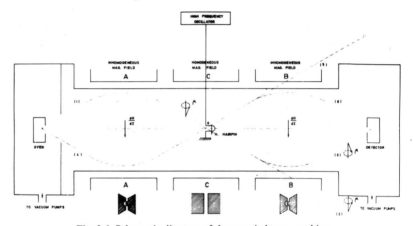

Fig. 5-6. Schematic diagram of the atomic beam machine.

In the O.D.T.Ü. machine the C field is adapted from the Brookhaven design (Cohen 1969) with slight modifications and consists of four Armco magnetic iron plates $\frac{1}{2}$ in. thick, 3 in. wide, and 14 in. long. The plates are separated from each other and from the main yoke members by quartz separators, the main gap being 0.550 in. wide. The quartz separators are produced in an optical shop to equal thickness of all pieces as good as 1 part in 10^4. This assures that the two inner plates of the magnetic field are parallel to each other to about the same accuracy. Distortions in the magnetic field arising from possible irregularities in the grain structure of the pole plates still remain. Nevertheless the homogeneity of the C magnet is expected to be appropriate for the most precise measurements. The Brookhaven design is modified also in such a way that the whole sandwich of four plates is separated electrically from the system in order to allow the use of the same magnet pole piece plates for electric field applications.

If the applied RF frequency in this C field region is equal to the Larmor precessional frequency of the atomic system for a given magnetic field H, the orientation of the magnetic moment will be reversed as we discussed before. In order to detect that a transition took place, the two inhomogeneous magnetic fields, known as the A and B fields, are placed on either side of the C field.

The A and B inhomogeneous magnetic fields are the same as at the University of California, Berkeley, atomic-beam machine and have $\frac{1}{2}$-in. radius male and female vanadium permendur pole pieces, 21 in. long. Cross sections of these pole pieces are shown at the lower part of Figure 5-6.

The gap boundaries nearly correspond to the equipotentials of the two-wire deflecting field made by Rabi, Kellogg, and Zacharias (1934) of the type first made by Millman, Rabi, and Zacharias. The magnetic field in the A and B regions is expected to be as high as 10,000 gauss and the magnetic field homogeneity about 5,000 gauss per cm.

Considering the solid angle of the beam, an atom of magnetic moment μ such as the one shown on the left side (upper side in the top view of the Figure 5-6) of the beam will be deflected to the right (downward in the figure) by the inhomogeneous magnetic field A. If this molecule does not undergo a transition, the direction of the force is still the same in the B region, and it will be thrown out to the side of the machine. If it undergoes a transition in the C region, however, because of the change of the direction of the magnetic moment it will be deflected to the left by the same amount in the B region and will arrive at the detector.

It should be noted that the deflection of the individual atoms is greatly exaggerated in Figure 5-6. Actual deflections are within a fraction of a millimeter. At exact resonance frequency an increase in the detector current is observed, and the frequency corresponding to this is recorded. To obtain the precise value of the magnetic field, the experiment is repeated alternately with a sample of known magnetic moment and the value of H is obtained from the corresponding resonance frequency.

Figure 5-7 shows the block diagram of the radiofrequency system. The oscillating field at the hairpin is produced by the klystron oscillator. Its frequency of oscillation is phase locked with a schomandl oscillator in such a way that a certain harmonic of the schomandl oscillator and the klystron output is fed to a mixer and the output of the mixer (which is the difference of these two frequencies) is sent to a synchriminator, designed to maintain this difference at 30 MHz to a high precision.

Any tendency of a shift in the klystron signal tending to change this 30-MHz difference is corrected by an error signal to the klystron oscillator from the synchriminator.

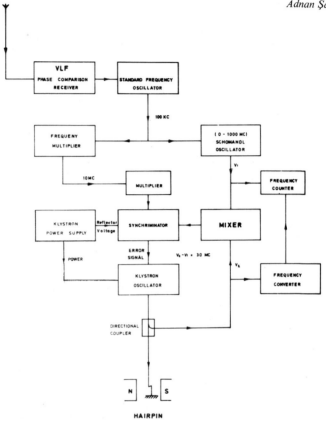

Fig. 5-7. Block diagram of the radiofrequency phase-locking system.

The schomandl oscillator in turn is phase locked to a 100 kHz signal from the standard frequency oscillator, which is accurate to several parts in 10^{10}.

Standard signals as good as several parts in 10^{12} are received by a very low frequency phase comparison receiver and compared with the standard frequency oscillator. Any shift beyond several parts in 10^9 in the frequency of the standard oscillator becomes known and is re-adjusted at regular intervals.

With the aid of this system, transition frequencies are obtained at the standard oscillator accuracy which is up to several parts in 10^{10}.

Figure 5-8 below shows how a moderate homogeneous magnetic field H and a simple hairpin system become a 2-ton machine.

The square sides which appear in the photograph are yoke members of the A and B magnets, and the semicylindrical cover plate of the C field

Fig. 5-8. General view of the O.D.T.Ü. atomic and molecular beam machine near completion.

is partly visible in the middle. The machine is symmetric about the C field, and the total distance from oven slit to the detector wire is $94\frac{1}{4}$ inches.

Acknowledgment

This work carried out in collaboration with the University of California, Berekeley, California, and Brookhaven National Laboratory, Atomic Beam Group, Brookhaven, New York, U.S.A.

My co-workers have been E. Aygün, T. İncesu, A. Özmen, and A. Tezer, Department of Physics, Orta-Doğu Teknik Üniversitesi, Ankara, Turkey.

REFERENCES

Bohr, A. 1951. *Phys. Rev.* 81:134, 331.
Bohr, A., and Weisskopf, V. F. 1950. *Phys. Rev.* 77:94.
Cohen, V. W. 1969. Private communication. Brookhaven National Laboratory, Upton, N.Y., U.S.A.
Cohen, V. W.; Nathans, R.; Silsbee, H.; Lipworth, E.; and Ramsey, R. 1968. The Electric Dipole Moment of the Neutron. International Conference on Atomic Physics, June 3–7. *Abstracts*, p. 8.
Dunoyer, L.
    1911a. *Compt. Rend.* Acad. Sci. Paris 152:594.
    1911b. *Le Radium* 8:142.
Fermi, E. 1930. *Z. Physik* 60:320.
Goudsmit, S., and Bacher, R.
    1929. *Phys. Rev.* 34:1499.
    1931. *Phys. Rev.* 37:663.
Kellogg, J. M. B.; Rabi, I. I.; Ramsey, N. F.; and Zacharias, J. R.
    1939a. *Phys. Rev.* 55:318.
    1939b. *Phys. Rev.* 56:728.
Kopfermann, H. 1940. *Kernmomenta*. Leipzig: Akademische Verlagsgesellschaft, M.B.H., and Ann Arbor, Michigan, U.S.A.: Edwards Bros., and new editions in German and English, 1955.
Kusch, P.; Millman, S.; and Rabi, I. I. 1940. *Phys. Rev.* 57:765.
Rabi, I. I.; Kellogg, J. M. B.; and Zacharias, J. R. 1934. *Phys. Rev.* 46:157.
Rabi, I. I.; Zacharias, J. R.; Millman, S.; and Kusch, P. 1938. *Phys. Rev.* 53:318.
Ramsey, N. F. 1953. Nuclear Moments. New York: Wiley.

6

RICHARD MARRUS

Recent Developments in Atomic Beams

This chapter will describe a few of the recent experiments employing the atomic beam method. Because of space limitations no attempt is made at a complete survey.

Historically, the atomic beam method has been useful in studying the energy levels of free atoms. In a beam, an atom is essentially free of all external perturbations, and so one can be sure that the observed spectra arise only from properties of the internal atomic Hamiltonian. Ever since the evolution of the magnetic resonance method by Rabi and collaborators, the main preoccupation has been with the study of hyperfine structure and the magnetic properties of atoms. From such studies, values of nuclear magnetic and electric quadrupole moments can be obtained, as well as the electronic g factor (g_J), the electronic and nuclear spins, and the hyperfine structure separations. From these studies several significant discoveries in both nuclear and atomic physics were made, notably the quadrupole moment of the deuteron, the anomalous electron g factor, the hyperfine structure of hydrogen, the Lamb shift, etc.

More recently, atomic beam experiments have been concerned with the study of atoms in an electric field, and it is some of these experiments that will be described here. However, for the most part, these experiments are rooted in the earlier hyperfine structure experiments, and the reader is referred to the book by Ramsey (1956) for background material and for references to earlier work.

THE PROTON–ELECTRON CHARGE DIFFERENCE

Perhaps the most fundamental of the atomic beam experiments employing electric fields has been carried out at Yale by Hughes and collaborators over a period of some twenty years (Zorn, Chamberlain, and Hughes 1963; Hughes 1964). Its purpose is to put the notion that the proton and electron charges are precisely equal and opposite on as firm an experimental foundation as possible.

There is ample motivation for such a program. Although accepted physical theory rests on the assumption of exact equality, there has been a great deal of speculation about the possible consequences of a small charge difference.

In fundamental particle physics, the conservation laws of charge, baryon number, and lepton number are believed to be absolute and to apply to all particle reactions. For example, the absence of the particle reaction $p^+ \rightarrow e^+ + \pi^0$ in nature can be explained by baryon conservation. However, if a small p^+–e^+ charge difference exists, then charge conservation could be used to explain its absence. More generally, the law of baryon conservation would follow from charge conservation instead of being an independent principle (Feinberg and Goldhaber 1959).

Lyttleton and Bondi (1959) have considered the possibility that the observed rate of expansion of the universe could be explained in terms of a small proton–electron charge difference. To get a feeling for the order of magnitude involved, consider the universe to be a homogeneous isotropic gas of H atoms. Then, if the proton charge is

$$q_p = (1 + y)e$$

where $-e$ is the electron charge, then there will be a Coulomb repulsion between any two H atoms given by

$$F_{\text{coul}} = \frac{(ye)^2}{r^2}$$

There will also be a gravitational attraction between them equal to

$$F_{\text{grav}} = G\frac{m_p^2}{r^2}$$

If the Coulomb repulsion is to be responsible for the observed expansion, then we must require

$$\frac{(ye)^2}{r^2} > G\frac{m_p^2}{r^2} \qquad \text{or} \qquad y > G^{\frac{1}{2}}\frac{m_p}{e} = 9 \times 10^{-19}$$

For agreement with the actual observed rate, Lyttleton and Bondi (1959) calculate

$$y \approx 2 \times 10^{-18} \tag{6-1}$$

A net charge difference would imply that rotating astronomical objects such as the earth would give rise to an approximately dipole magnetic field whose axis is along the rotational axis. Since this is a not unreasonable picture of the earth's field, it is of interest to ask what the magnitude of the earth's field implies for y. A simple calculation shows that the field at the equator is

$$B = 0.8\frac{\rho_M}{M}N_0\frac{e}{c}Zy\omega R_e^2 \tag{6-2}$$

where ρ_M = mean mass density of earth ≈ 5.5 gm per cm^2; M = mean atomic weight of earth atom; Z = mean atomic number of earth atoms $(Z/M \approx \frac{1}{2})$; N_0 = Avogadro's number; ω = angular velocity of earth; R_e = earth's radius. For a field $B = 0.2$ gauss, we require $y \approx 3 \times 10^{-19}$.

The atomic beam experiment of Hughes and collaborators is based on the straightforward idea that a charged atom will experience a force $F = ZyeE$ if placed in a homogeneous field E and hence will be deflected. A diagram of the Hughes apparatus is shown in Figure 6-1. Clearly, for maximum sensitivity, good collimation, a long apparatus, intense electric fields, and a sensitive technique for measuring beam deflections are desirable. In the most recent report (Fraser, Carlson, Hughes 1968), overall beam length is 7.2 m, and electric fields up to 200 kV per cm are obtained. The method of measuring the size of the deflection is of some interest because similar techniques have been used before and since to measure small frequency shifts in resonance experiments. With no electric field present, the intensity pattern is traced out (Fig. 6-2). The

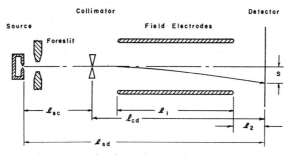

Fig. 6-1. Geometry of apparatus, showing trajectory of an atom which has been deflected by the electric field. Dimensions (in centimeters) are:

	Alkali experiment	Hydrogen experiment
ℓ_1	200	94
ℓ_2	30	14.6
ℓ_{sc}	193	41
ℓ_{sd}	455	163

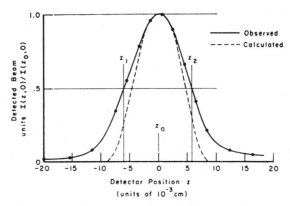

Fig. 6-2. Theoretical beam shape (calculated on the basis of classical trajectories and ideal geometry) compared to an observed $I(z, 0)$ for K atoms.

detector is then placed at the position Z_1 where the slope of the pattern $(\partial I/\partial Z)$ is a maximum. The electric field is then turned on, and the change in intensity δI is then measured. Making some reasonable assumptions about the velocity distribution, it can be shown that

$$\delta I = \left(\frac{\partial I}{\partial Z}\right) S_\alpha \tag{6-3}$$

where S_α is the deflection of an atom in the beam having the most probable velocity. The situation here is very similar to the triode amplifier, where the tube characteristic is exploited to generate a large plate signal from a small grid signal. Here, a small deflection can give rise to a large intensity change by using a sharply sloping intensity pattern. With this scheme, deflections of order 10^{-5} cm could be detected, and it was determined that

$$|q(Cs)| \leqslant 1.7 \times 10^{-18}e \qquad |q(K)| \leqslant 1.3 \times 10^{-18}e$$

Writing that

$$q(Cs) = 55\,\delta q + 78\,q_n$$
$$q(K) = 19\,\delta q + 20\,q_n$$

where δq = electron–proton charge difference and q_n = neutron charge, these results give

$$|\delta q| \leqslant 3.5 \times 10^{-19}e \qquad |q_n| \leqslant 2.7 \times 10^{-19}e$$

Using the beta decay reaction $n \to p^+ + e^- + \bar{\nu}$, and assuming charge conservation and $q(\bar{v}) = 0$, then it would follow that $\delta q = q_n$. With this assumption,

$$|\delta q| = |q_n| \leqslant 1.3 \times 10^{-20} e$$

ATOMIC BEAM METHOD
FOR STUDYING OPTICAL STARK EFFECT AND ISOTOPE SHIFTS

A new method for studying these parameters was developed in 1965 by Marrus and McColm (1965) at Berkeley and subsequently employed in the study of the Stark effect in the alkali resonance lines (Marrus, McColm, and Yellin 1966; Marrus and Yellin 1969) and the isotope shifts in the radioactive cesium isotopes (Marrus, Wang, and Yellin 1967, 1969). The method employs an atomic beam apparatus of the flop-in type (Zacharias 1942; see Fig. 6-3). The basic idea of such an apparatus is that if one succeeds in flipping the spin of the atom in the C region then an atom is refocused and strikes the detector. If the spin direction is unchanged when the atom leaves the C region, then the atom goes into the magnet walls and never reaches the detector.

This method of detection gives the atomic beam method an important advantage over optical methods in studying radioactive isotopes. With the optical methods, detection is always on the light. This implies that the small number of radioactive atoms created in the production process must be separated from the large number of carrier atoms; otherwise the weak light of the small number of radioactive atoms will be swamped by the intense light of the carrier atoms. Therefore, in the case of neutron-produced isotopes, an expensive and time-consuming mass separation must be done. With cyclotron-produced isotopes, a chemistry must be done to separate them. Both of these processes put lower limits on the half-lives of radioisotopes than can be studied. In addition, optical

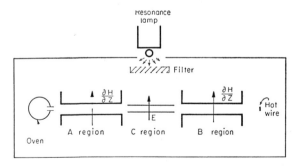

Fig. 6-3. Schematic diagram of atomic beam apparatus for studying Stark effect.

methods have the problem that interaction of the microscopic number of radioactive atoms with the walls of their container can quickly destroy the sample's usefulness. Although some progress has been made in applying optical techniques to radioisotopes (Davis, Aung, Kleinman 1966; Hühnermann and Wagner 1966, 1967, 1968), optical techniques have not had the general success of atomic beam techniques.

Atomic beam experiments escape the problems described above by detecting directly on the radioactivity of the atoms being studied. The detector in an atomic beam experiment consists of a metal foil chosen so that a beam atom striking the surface will stick to it with a high probability. The foil is exposed to the beam for some predetermined time period and is then placed in a suitable radiation counter to measure the radioactivity deposited. The presence of stable atoms on the foil is clearly of no significance. Moreover, the wall problem is not relevant to an atomic beam experiment. The success of the atomic beam method is reflected in the fact that successful hyperfine structure (hfs) experiments have been done on several hundred radioisotopes.

The basic apparatus is a conventional atomic beam apparatus with flop-in magnet geometry and is shown in Figure 6-3. However, the C region consists of a pair of electric field plates capable of sustaining fields of almost 10^6 volts per cm. The region between the plates is illuminated by resonance radiation from a ^{133}Cs resonance lamp placed outside the apparatus. The light is filtered so that only the D_1 transition is passed. The D_1 line consists of two components (Fig. 6-4) separated by the hfs of the ground state. The hfs of the excited $6^2P_{\frac{1}{2}}$ state is much

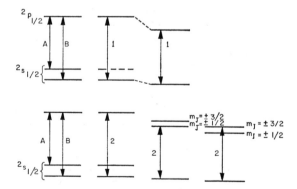

Fig. 6-4. Schematic diagram of energy levels. The lines A and B are both present in the lamp. At zero electric field the absorption lines 1 and 2 coincide with the emission line B. Signals are also observed at electric fields such that the lines 1 and 2 are made to resonate with the line A.

smaller than the ground state and is incompletely resolved in the lamp. Consider now an atomic beam of ^{133}Cs in the apparatus with zero electric field ($E = 0$). Under this condition, the absorption lines of atoms in the beam coincide with the emission lines of atoms in the lamp, and some of the atoms will be excited to $6^2P_{\frac{1}{2}}$ by resonant absorption of optical photons. However, the lifetime of this state is short ($\sim 10^{-8}$ sec) compared to the transit time down the apparatus, and half the atoms decay to the ground state so that their spin is flipped. These atoms are refocused on the detector and give rise to a large signal.

The Stark effect can now be determined by applying an electric field. As will be seen shortly, the effect of an electric field on a state with $J = \frac{1}{2}$ (the two levels involved in the D_1 transition) is especially simple. To the accuracy of these experiments, all of the hyperfine levels associated with the state are shifted by an equal amount. Relative shifts of the hyperfine levels and lifting of the Zeeman degeneracy are negligibly small. Hence an applied electric field does not split the levels but decreases the energy of the two absorption lines as shown in Figure 6-4. When the energy decreases sufficiently, the absorption lines no longer are coincident with the lamp emission lines, and the flop-in signal goes to zero. However, at a sufficiently high field, the energy is decreased to the point where absorption line 1 of atoms in the beam overlaps emission line A of the lamp. At this point a new signal increase is observed which indicates that the electric field corresponds to an energy shift equal to the ground state hfs. From the assumption that the energy shift (W) is proportional to the square of the electric field (E^2), this enables an experimental determination of the Stark effect to be made in the D_1 line. The observed signal is shown in Figure 6-5a.

A more complete study of the Stark effect in the 6P state can be made by extending the measurements to the D_2 line, i.e., the transition $6^2P_{\frac{3}{2}}-6^2S_{\frac{1}{2}}$. In analyzing the Stark effect of beam atoms as the electric field is applied, we can no longer make the simplifying assumption that the electric field shifts all the hyperfine levels of the excited states by equal amounts, since we now have $J = \frac{3}{2}$. However, a new simplifying assumption is possible because the induced Stark shifts of 9,192 MHz (i.e., the ground state hfs) is fifteen times greater than the overall hyperfine width of $6^2P_{\frac{3}{2}}$. In this situation, it is possible to treat the hyperfine Hamiltonian ($\mathscr{H}^{hfs}$) as a first-order perturbation on the Stark Hamiltonian. If the electric field is in the Z direction, then J_Z and I_Z will commute with the Stark Hamiltonian

$$\mathscr{H}^E = -\bar{p} \cdot \bar{E} \qquad \text{with} \qquad \bar{p} = e\bar{r}$$

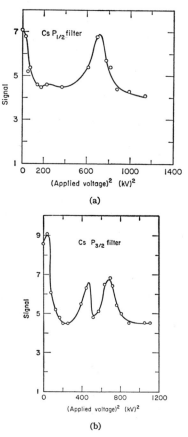

(a)

(b)

Fig. 6-5. (a) Observed cesium signal with D_1 radiation only. (b) Observed cesium signal with D_2 radiation only.

and the first-order effect of $\mathcal{H}^{hfs}$ can be considered in an (m_J, m_I) representation. A simple calculation shows that the effect of $\mathcal{H}^{hfs}$ is to broaden a Stark-shifted line by an amount of order the hfs of the $6\,^2P_{\frac{3}{2}}$ state. Since this is small compared to the line width of the signals (about 1500 MHz, determined by the width of the lamp), it suggests that we can treat the Stark effect by first ignoring hfs and then assuming that the hfs merely gives rise to a width in the calculated intensity pattern.

The resulting energy level pattern is shown in Figure 6-4. On very general grounds, the Stark effect splits $6\,^2P_{\frac{3}{2}}$ into a doublet characterized by quantum numbers $m_J = \pm\frac{3}{2}$ and $m_J = \pm\frac{1}{2}$. Hence, overlaps will occur when absorption lines 1 and 2 overlap lamp line A. No other signals are observed because the design of the apparatus precludes the observation of signals from the upper hyperfine state. The observed signal pattern is shown in Figure 6-5b and is in agreement with this model.

The positions of the two high-field peaks associated with D_2 and the single peak associated with D_1 are sufficient to determine the Stark effect of the 6P state. The results can be analyzed in terms of the polarizabilities $\alpha(n^2 l_j m_j)$ associated with the state and defined by the equation

$$W(n^2 l_j m_j) = -\tfrac{1}{2}\alpha(n^2 l_j m_j)E^2$$

The equations expressing the three observed peaks become

$$-\tfrac{1}{2}E_1^2[\alpha(6\,{}^2P_{\frac{1}{2}} \pm \tfrac{1}{2}) - \alpha(6\,{}^2S_{\frac{1}{2}} \pm \tfrac{1}{2})] = 9192 \text{ MHz}$$

$$-\tfrac{1}{2}E_2^2[\alpha(6\,{}^2P_{\frac{3}{2}} \pm \tfrac{3}{2}) - \alpha(6\,{}^2S_{\frac{1}{2}} \pm \tfrac{1}{2})] = 9192 \text{ MHz}$$

$$-\tfrac{1}{2}E_3^2[\alpha(6\,{}^2P_{\frac{3}{2}} \pm \tfrac{1}{2}) - \alpha(6\,{}^2S_{\frac{1}{2}} \pm \tfrac{1}{2})] = 9192 \text{ MHz}$$

where E_1, E_2, and E_3 are the measured electric fields at which the peaks occur. Although there are four polarizabilities and three equations, the ground state polarizabilities $\alpha(6\,{}^2S_{\frac{1}{2}} \pm \tfrac{1}{2})$ have been measured in an independent experiment by Salop, Pollack, and Bederson (1961). Hence these polarizabilities can be used to determine the 6P polarizabilities. Results for K, Rb, and Cs are given in Table 6-1 and compared with theoretical predictions of the Bates–Daamgard method. Agreement is seen to be good.

The extension of the method to isotope shift (IS) measurements is, in principle, straightforward. One continues to use a ^{133}Cs lamp, but another Cs isotope is used in the beam. Because of the different ground state hfs and the isotope shift, there will, in general, be no signal at zero

Table 6-1. Alkali polarizabilities $\times\ 10^{24}$ cm^3

Atom		${}^2S_{\frac{1}{2}}$	${}^2P_{\frac{1}{2}}$	${}^2P_{\frac{3}{2}} \pm \tfrac{3}{2}$	${}^2P_{\frac{3}{2}} \pm \tfrac{1}{2}$
Cs	Experiment:	52.5(6.5)[a]	187(29)	196(30)	273(42)
	Theory[b]:	56	192	191	246
Rb	Experiment:	40(5)[a]	112(17)	102(15)	148(23)
	Theory[b]:	46	116	108	151
K	Experiment:	36(4.5)[a]	87(13)	68(10)	114(16)
	Theory[b]:	41	93	80	109

[a] Salop, Pollack, and Bederson (1961).
[b] Calculated using the theoretical oscillator strength of D. R. Bates and A. Daamgard (1949).

electric field. However, by application of a suitable field, resonances can be observed. By using the Stark effect calibration described in the previous experiment, the positions of these resonances determine the isotope shifts. However, the method as it has been described so far is not of sufficient precision to give quantitatively useful information. A straightforward calculation shows that the IS predicted for the cesium isotopes, assuming that the nuclear radius (R), goes as

$$R = R_0 A^{\frac{1}{3}} \qquad (R_0 = 1.2 \times 10^{-13} \text{ cm})$$

gives 300 MHz per neutron. Now isotope shifts tend to be smaller than predicted by this model so that the line width of the experiment should be small compared to this. There are two important sources of line width:

(a) Electric field inhomogeneities: an electric field inhomogeneity ΔE will produce an energy broadening ΔW

$$\frac{\Delta W}{W} = \frac{2\Delta E}{E}$$

At a field so that $W = 9192$ MHz, we would like ΔW to be less than the natural width of 10 MHz, so we would like to have

$$\frac{\Delta E}{E} < \frac{1}{2} \times 10^{-3}$$

With the intense fields required in these experiments, the best homo-geneity achieved to date is $\sim \frac{1}{2} \times 10^{-2}$.

(b) Line width of the lamp: the line width of the lamp is primarily the result of Doppler broadening and is about 1500 MHz. In order to improve on this, a technique that has been used is to employ a lamp of the beam type. Although the line width of this type of lamp can be extremely small, the intensity is very weak. Since the signal in these experiments is proportional to lamp intensity, what is desired is the line width of the beam lamp but the intensity of an RF discharge lamp. This is achieved by using an RF discharge lamp and passing the D_1 light from the lamp through an optically dense absorption beam. The effect of the absorption beam is to remove from the optical beam those photons which lie within the absorption width (~ 150 MHz) of beam atoms. Hence the spectral profile of the light incident on the atomic beam apparatus has no photons at frequencies corresponding to an absorption line in ^{133}Cs. Therefore, the signal intensity plotted against E^2 will have minima at positions where an absorption line of atoms in the atomic beam apparatus overlaps an absorption line of atoms in the absorption beam. These minima are clearly indicated in Figure 6-6. It is seen that the width associated with the minima is about 150 MHz. Moreover, the hfs of the $6\,^2P_{\frac{1}{2}}$ state is clearly visible and can be measured to 40 MHz.

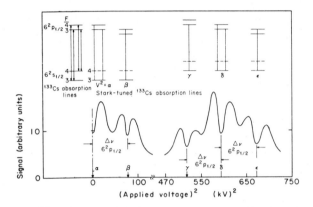

Fig. 6-6. Observed ^{133}Cs signal versus square of applied voltage. The position of the Stark-shifted absorption lines relative to those in the absorption cell for each of the observed minima is indicated directly above the minima. The separation between α and δ corresponds to a shift equal to the ground-state hyperfine separation and serves as a calibration.

With this improvement in precision, isotope shifts can be measured. A signal pattern associated with ^{134m}Cs ($\tau_{\frac{1}{2}} = 2.8$ hr) is shown in Figure 6-7. From this pattern the isotope shift is determined to be IS $= -2.2(1.2)$

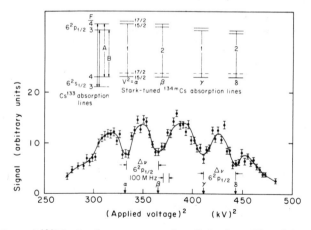

Fig. 6-7. Observed ^{134m}Cs signal versus square of applied voltage. The minimum α occurs when beam absorption line 1 coincides with ^{133}Cs absorption line A; β occurs when line 2 coincides with A; γ occurs when 1 coincides with B; and δ occurs when 2 coincides with B.

$\times\ 10^{-3}\ \mathrm{cm}^{-1}$. Similar experiments have been performed on several other radioisotopes of Cs with the results given in Table 6-2. The most striking feature is the smallness of the isotope shifts. A qualitative explanation for the size of the shifts may lie in the fact that ^{137}Cs has a magic number of neutrons ($N = 82$) so that it is highly spherical. Measurements of the quadrupole moments of ^{131}Cs and ^{132}Cs indicate that these nuclei are substantially deformed. It may be, therefore, that in the neutron-deficient Cs isotopes the deformation effect largely cancels the normal volume effect.

OTHER STARK-EFFECT EXPERIMENTS

In addition to the method just described for studying Stark effect, there are two other atomic beam methods that have been used to study the Stark effect.

A. A classical magnetic resonance experiment is done and a radio-frequency transition is observed. An electric field is then applied to this atom, and the shift in resonance frequency is measured. From this, the difference in polarizability of two Zeeman sublevels can be obtained.

B. The force on the induced dipole moment in the presence of an inhomogeneous electric field is studied. In the most straightforward application of this method, the deflection produced by the force $\bar{F} = -\bar{\nabla}(\bar{p} \cdot \bar{E})$ is measured. In a more ingenious variation of this method, the electric force is balanced by a known magnetic force on the magnetic dipole moment, and the polarizability is deduced from the known $\bar{E}$ and $\bar{B}$ fields at balance. Methods employing the force measure the absolute polarizability of the single level involved and are therefore complementary to the other methods.

The first experiment (Haun and Zacharias 1957) of Type A was designed to study the relative shift of the hyperfine levels in the ground state of ^{133}Cs when an electric field is applied. The second-order energy shift to the ground state of Cs in the presence of an electric field is given by the well-known matrix element

$$W = \sum_{\gamma} \frac{|\langle\gamma| -e\bar{r} \cdot \bar{E} |6\ ^{2}\mathrm{S}_{\frac{1}{2}} Fm_{F}\rangle|^{2}}{\Delta W(\gamma; Fm_{F})}$$

As mentioned earlier, the matrix element squared turns out to be independent of all the angular momenta. Therefore, the only contribution

Table 6-2. Measured isotope shifts in the cesium isotopes (10^{-3} cm^{-1})

Isotope:	127	129	131	132	134	134m	135	137
Atomic beam[a]	5.9(1.5)	2.8(1.5)	—	—	1.8(1.0)	−2.2(1.2)	—	−6.0(1.5)
Otten and Ullrich[b]	—	—	−0.31(5)	1.6(5)	—	—	—	—
Hühnermann and Wagner[c]	—	—	0.39(9)	—	1.17(5)	—	−1.23(7)	−4.81(6)

[a] Marrus, Wang, and Yellin (1967, 1969).
[b] Otten and Ullrich.
[c] Hühnermann and Wagner (1966, 1967, 1968).

to a relative shift of the two hyperfine levels $F = 3$ and $F = 4$ can come from the difference in the energy denominator of the two levels. Such a shift is obviously quite small and is of order of magnitude

$$0\left(\frac{\text{hfs separation}}{\text{optical energy}}\right) \approx 10^{-5}$$

To study the effect, an observation was made of the field-independent transition $(F = 4; m_F = 0) \leftrightarrow (F = 3; m_F = 0)$ at low-magnitude fields (<0.5 gauss). The resonance was traced out with the electric field on and the electric field off and the shift directly measured. Line width in the experiment was about 120 cps. The result obtained was

$$\delta v = -2.29 \times 10^{-6}(1 \pm 0.03)E^2 \text{ cps}$$

with E in volts per cm. More recently, a very similar experiment (Snider 1966) has been made on the ^{39}K ground state and a hydrogen maser experiment (Fortson, Kleppner, and Ramsey 1964) for the H ground state. The values found for ^{39}K and for H are, respectively:

$$^{39}\text{K}: \qquad \delta v = -(7.60 \pm 0.76) \times 10^{-8} E^2 \text{ cps}$$

$$\text{H}: \qquad \delta v = -(7.4 \pm 0.5) \times 10^{-10} E^2 \text{ cps}$$

Anderson (1961) has devised a theory to explain these results. He notes that not only is it important to consider differences in the energy denominator resulting from hyperfine structure but that corrections to the wavefunction arising from hfs also contribute significantly. Anderson's theory yields for these experiments:

$$\text{Cs (6S)}: \qquad \delta v = -2.11 \times 10^{-6} E^2 \text{ cps}$$

$$\text{K (4S)}: \qquad \delta v = -6.27 \times 10^{-8} E^2 \text{ cps}$$

$$\text{H (1S)}: \qquad \delta v = -7.3 \times 10^{-10} E^2 \text{ cps}$$

Sandars (1967) has recently redone the hydrogen case by a new method and finds

$$\delta v = -8.7 \times 10^{-10} E^2$$

Experiments which study the polarizabilities of atomic levels by measuring the force on the atom in an inhomogeneous electric field have been done at Yale (Chamberlain and Zorn 1963), at Michigan (Hall and Zorn 1967; Hall 1967), and at New York University by Salop, Pollack, and Bederson (1961).

The experiments of Zorn are very similar in method to the charge difference experiment described in the first section of this chapter. The apparatus of Figure 6-1 is identical to Zorn's, but the homogeneous electric field is replaced by an inhomogeneous electric field characterized by a gradient $\partial E/\partial Z$. In the presence of such a field, an atom will experience a force (F)

$$F_{elect} = \alpha E \frac{\partial E}{\partial Z} \tag{6-4}$$

where α is the polarizability. The fields used in these experiments are those characteristic of the two-wire field. An important property of the two-wire field is that at a distance of $1.2a$ along the perpendicular to the line joining the two wires (separation between the wires is $2a$) the field and field gradient are relatively constant. Hence, by good beam collimation, a uniform field and field gradient can be applied to the entire beam along its trajectory.

The deflection can be determined according to Equation 6-3 and easily related to the polarizability. A criticism of Zorn's original work (Chamberlain and Zorn 1963) was that the measured deflection depended on assumptions about the velocity distribution. More recently Zorn and Hall (1967; Hall 1967) have repeated the measurements on the ground state polarizabilities using a velocity selector and better vacuum. The results are somewhat higher and are shown in Table 6-3 along with values obtained by the *E–H* gradient balance method and theoretical measurements.

The *E–H* gradient balance method avoids the difficulty of velocity-dependent deflections by balancing the applied electric force by an equal

Table 6-3. Ground state polarizabilities of alkali atoms ($\times 10^{24}$ cm^3)

Method	Na	K	Rb	Cs
E-H gradient balance[a]	20(2.5)	36.5(4.5)	40(5)	52.5(6.5)
Deflection, no velocity selection[b]	21.5(2)	38(4)	38(4)	48(6)
Deflection, velocity selection[c]	24.4(1.7)	45.2(3.2)	48.7(3.4)	63.3(4.6)
Theory[d]	24.6(1.2)	41.6(2.1)	43.8(2.2)	53.7(5.4)
Theory[e]	22.9	44.4	49.1	67.7
Theory[f]	23	43	46	56

[a] Salop, Pollack, and Bederson (1961).
[b] Chamberlain and Zorn (1963).
[c] Hall and Zorn (1967).
[d] Dalgarno and Kingston.
[e] Sternheimer.
[f] Bates and Daamgard.

and opposite magnetic force. In the presence of an inhomogeneous magnetic field characterized by field B and field gradient $\partial B / \partial Z$, there will be a magnetic force given by

$$F_{\text{mag}} = \mu_{\text{eff}} \frac{\partial B}{\partial Z} \tag{6-5}$$

where μ_{eff} is the effective magnetic moment of the atom in the field B and is defined by

$$\mu_{\text{eff}} = -\frac{\partial W_{\text{mag}}}{\partial B} \tag{6-6}$$

where W_{mag} is the magnetic energy of the atom. Hence, at balance we can write

$$F_{\text{elect}} = F_{\text{mag}}$$

or

$$\alpha E \frac{\partial E}{\partial Z} = \mu_{\text{eff}} \frac{\partial B}{\partial Z} \tag{6-7}$$

This balance condition is velocity independent and hence avoids assumptions about the velocity distribution of atoms in the beam.

The electric and magnetic fields in the experiment are produced by cutting two magnetic pole pieces in the two-wire geometry and electrically insulating them from each other so that a voltage may be applied across the gap. Because of the congruent geometry, it is possible to write that

$$\frac{1}{E} \frac{\partial E}{\partial Z} = \frac{1}{B} \frac{\partial B}{\partial Z}$$

Hence Equation 6-7 becomes

$$\alpha = \frac{\mu_{\text{eff}} B}{E^2} \tag{6-8}$$

The parameters μ_{eff} and B are easily measurable, and E can be determined from the applied voltage and the known properties of the two-wire field.

Figure 6-8 shows the schematic of the atomic beam apparatus used. The balance condition can be satisfied only for beams of infinitesimal width, and it is therefore important to have good beam collimation. The oven slit is 0.0025 in. wide, the collimator slit is 0.002 in. wide, and the hot wire detector is 0.002 in. in diameter, able to be positioned to within 0.0005 in. The pole pieces fit snugly inside a rectangular glass tube which

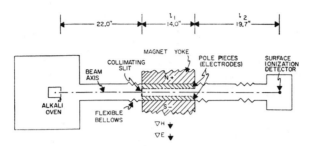

Fig. 6-8. Schematic diagram of the atomic beam apparatus.

acts as a vacuum envelope. The magnet yoke is outside the vacuum system. The glass envelope serves the essential function of electrically insulating the pole pieces from each other so that a potential difference can be maintained.

Figure 6-9 shows a qualitative demonstration of the congruency of the *E–B* fields on a potassium beam. Curve A shows a beam profile (intensity vs. detector position) with no fields present, i.e., $B = E = 0$. Curve B shows a profile with $E = 0$ but $B = 317$ gauss. This field is sufficient to decouple the electronic and nuclear spins, and the profile clearly exhibits the two peaks corresponding to $\mu_{\text{eff}} = \pm \mu_0$. This is the typical Stern–Gerlach pattern for an electronic magnetic moment associated with a $^2S_{\frac{1}{2}}$ state. The third profile, case C, corresponds to the case of the electric field tuned to one of the balance positions, that corre-

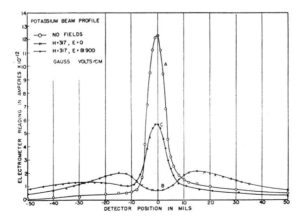

Fig. 6-9. Qualitative demonstration of E-H field congruency, showing potassium beam profile under the conditions: A, no fields; B, magnetic field and no electric field; C, magnetic and electric fields at balance condition.

sponding to $\mu_{\text{eff}} = -\mu_0$. The left-hand peak of pattern B, that corresponding to $\mu_{\text{eff}} = +\mu_0$, is shifted substantially to the left by the electrostatic deflecting force and further broadened by the velocity distribution. However, that corresponding to $\mu_{\text{eff}} = -\mu_0$ has been brought back to the undeflected position of peak A and the width closely approximates that of the original beam shape.

In the case of Cs, the large hfs made it inconvenient to try to decouple μ_I and μ_J, and the actual method for determining the polarizability can be understood with reference to Figure 6-10. This is a plot of effective magnetic moment μ/μ_0 vs. the applied field B. With the electric field $E = 0$, the magnetic field is set so that one of the states has zero moment and passes undeflected down the apparatus. This serves to determine B. The electric field is then turned on, and this has the effect of bringing one by one the states with negative μ_{eff} into the balance condition. At the position of the second zero moment field as indicated here, it should be possible to observe seven balance peaks. However, because of limits imposed by electric field breakdown, only three of the peaks could be seen. Results for the polarizabilities obtained here can be checked by looking also with B set for the third zero moment position. Results obtained by this method are shown in Table 6-3 where they can be compared with other experimental and theoretical values for the alkali polarizabilities.

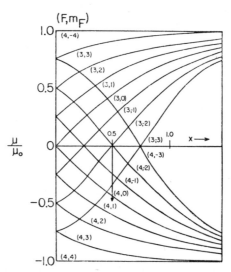

Fig. 6-10. The variation of effective magnetic moments with dimensionless parameter x (proportional to H) for an $I = \frac{7}{2}$, $J = \frac{1}{2}$ atom (^{133}Cs).

REFERENCES

Anderson, L. W. 1961. *Nuovo Cimento* 22:936.

Bates, D. R., and Daamgard, A. 1949. *Phil. Trans. Roy. Soc.* London A242:101.

Chamberlain, G. C., and Zorn, J. C. 1963. *Phys. Rev.* 129:677.

Davis, S. P.; Aung, T.; and Kleinman, H. 1966. *Phys. Rev.* 147:861.

Feinberg, G., and Goldhaber, M. 1959. *Proc. Natl. Acad. Sci. U.S.* 45:1301.

Fortson, E. W.; Kleppner, D.; and Ramsey, N. F. 1964. *Phys. Rev. Letters* 13:22.

Fraser, L. J.; Carlson, E. R.; and Hughes, V. W. 1968. *Bull. Am. Phys. Soc.* 13:636.

Hall, W. D. 1967. Thesis, University of Michigan, Ann Arbor, Michigan.

Hall, W. D., and Zorn, J. C. 1967. *Bull. Am. Phys. Soc.* 12:131.

Haun, R. D., Jr., and Zacharias, J. R. 1957. *Phys. Rev.* 107:107.

Hühnermann, H., and Wagner, H.
   1966. *Phys. Rev. Letters* 21:303.
   1967. *Z. Physik* 199:239.
   1968. *Z. Physik* 216:38.

Hughes, V. W. 1964. Page 6 in *Gravitation and Relativity*, ed. H. Y. Chiu and W. F. Hoffmann. New York and Amsterdam: W. A. Benjamin.

Lyttleton, R. A., and Bondi, H. 1959. *Proc. Roy. Soc.* London A252:313.

Marrus, R., and McColm, D. W. 1965. *Phys. Rev. Letters* 15:813.

Marrus, R.; McColm, D.; and Yellin, J. 1966. *Phys. Rev.* 147:55.

Marrus, R.; Wang, E.; and Yellin, J.
   1967. *Phys. Rev. Letters* 19:1.
   1969. *Phys. Rev.* 177:122.

Marrus, R., and Yellin, J. 1969. *Phys. Rev.* 177:127.

Otten, E. W., and Ullrich, S. Private communication to be published: *Erstes Physikalisches Institut der Universität Heidelberg.*

Ramsey, N. F. 1956. *Molecular Beams.* London, U.K.: Oxford University Press.

Salop, A.; Pollack, E.; and Bederson, B. 1961. *Phys. Rev.* 124:1431.

Sandars, P. G. H. 1967. *Proc. Roy. Soc.* London 92:857.

Snider, J. L. 1966. *Phys. Rev. Letters* 21:172.

Zacharias, J. R. 1942. *Phys. Rev.* 61:270.

Zorn, J. C.; Chamberlain, G. E.; and Hughes, V. W. 1963. *Phys. Rev.* 129:2566.

7

R. H. GARSTANG

Atomic Physics in Astrophysics

1. EXCITATION PROCESSES IN THE SOLAR CORONA

The solar corona has been known since ancient times. It is seen around the limb of the sun during total solar eclipses. The spectrum of the corona was first obtained at the eclipse of 1869, when the green coronal line at wavelength 5303 Å was discovered. The spectrum has been observed at many eclipses since that time, and more recently it has been observed without an eclipse by means of coronographs. A number of coronal lines are now known in the visible spectrum. With the advent of rocket spectroscopy observations of the coronal ultraviolet spectrum became possible, and a large number of additional lines were discovered.

The identification of the coronal lines in the visible spectrum was a difficult problem and was not solved until 1941. Edlén (the most accessible account of his work is Edlén 1945) showed that the lines were forbidden transitions in highly ionized atoms of iron and nickel. A few representative examples are listed in Table 7-1; a more extensive list may be found in Billings (1966, pp. 304–05). The high degree of ionization found for the coronal ions established that the corona is at a temperature of $10^{6\circ}$K. (There is much independent evidence which confirms this interpretation.)

Table 7-1. Some coronal lines

Ion	Wavelength (Å)	Transition
Fe XIV	5303	$3s^2 3p\ ^2P_{\frac{3}{2}} \rightarrow\ ^2P_{\frac{1}{2}}$
Fe X	6374	$3s^2 3p^5\ ^2P_{\frac{1}{2}} \rightarrow\ ^2P_{\frac{3}{2}}$
Ca XV	5694	$2p^2\ ^3P_1 \rightarrow\ ^3P_0$
Fe XV	7059	$3s3p\ ^3P_2 \rightarrow\ ^3P_1$
Fe XIII	10747	$3p^2\ ^3P_1 \rightarrow\ ^3P_0$
Fe XIII	10798	$3p^2\ ^3P_2 \rightarrow\ ^3P_1$
Fe XIII	3388	$3p^2\ ^1D_2 \rightarrow\ ^3P_2$
Fe XV	284.2	$3s3p\ ^1P_1 \rightarrow 3s^2\ ^1S_0$
Fe XV	417	$3s3p\ ^3P_1 \rightarrow 3s^2\ ^1S_0$
Fe XVII	16.77	$2p^5 3s\ ^1P_1 \rightarrow 2p^6\ ^1S_0$
Fe XXV	1.87	$1s2p\ ^1P_1 \rightarrow 1s^2\ ^1S_0$

The coronal gas is highly ionized so that there are many free electrons. The electron density decreases sharply outward from the solar limb; a typical value in the solar inner corona (where the coronal emission lines are most easily seen) is about 10^8 electrons per cm^3. It was soon realized that these electrons (with energies of the order of 100 eV) were responsible for exciting the coronal lines. In Fe XIV, which has a lowest electron configuration $3s^2 3p$ and term ^2P, the ions are excited from the $^2P_{\frac{1}{2}}$ to $^2P_{\frac{3}{2}}$ levels by electron excitation, and then the ions spontaneously emit the 5303 Å line, returning to the ground state (Fig. 7-1). Calculations show that excitation by absorption of solar radiation and deexcitation by collisions are negligible.

In this work it was tacitly assumed that higher energy levels play no part in the excitation of the 5303 Å line. After ultraviolet rocket observations had shown the existence of many strong ultraviolet coronal emission lines, all permitted transitions from high energy states, it was realized that these transitions may contribute to the population of the upper state of the 5303 Å line and hence to the intensity of the line. The problem was worked out by Pecker and Thomas (1962). They included in their calculations excitations and deexcitations for all levels in the $3s^2 3p$ and $3s 3p^2$ configurations (Fig. 7-1) in Fe XIV. As expected, their results showed that the

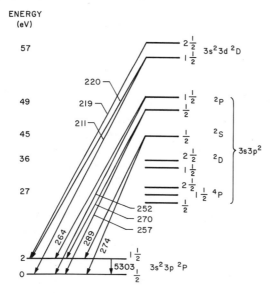

Fig. 7-1. Energy levels in Fe XIV, not to scale. All the levels of the $3s^2 3p$, $3s 3p^2$, and $3s^2 3d$ configurations are shown. The visible coronal line 5303 Å is indicated as well as lines identified in the ultraviolet spectrum. The energies (in electron volts) are indicated on the left.

ratio of the population of the $3s^23p$ $^2P_{\frac{3}{2}}$ state to that of the $3s^23p$ $^2P_{\frac{1}{2}}$ state is increased when allowance is made for the $3s3p^2$ configuration, by factors ranging from 2% at an electron density of 10^7 cm^{-3} to a factor of nearly 2 at an electron density of 10^{10} cm^{-3}. Pecker and Thomas also performed calculations on Fe X allowing for the excited configurations $3s3p^6$ and $3s^23p^43d$ and obtained a result similar to that for Fe XIV.

In calculations of atomic parameters of the types needed in work on coronal-line intensities care must be taken to allow for configuration interaction where this is significant. Garstang (1962) studied Si X, Fe X, and Fe XIV. In the case of Fe XIV he found configuration interaction between the terms $3s^23d$ ^2D and $3s3p^2$ ^2D to be very large. The $3s^23d$ ^2D term was unknown in the laboratory; its position had to be predicted by extrapolation. Transition probabilities were computed for all the allowed and forbidden transitions within and between the $3s^23p$, $3s^23d$, and $3s3p^2$ configurations. It was of course necessary to work in intermediate coupling because spin-orbit effects are large. Many individual lines were found to have line strengths which differed substantially from those calculated without the inclusion of configuration interaction. The importance of the interaction of the $3s^23d$ and $3s3p^2$ configurations is a good example of the importance of the concept of a complex for highly ionized atoms. (A *complex* is a set of electron configurations in an atom with a define parity and a prescribed set of principal quantum numbers, in this case three electrons with $n = 3$ and any allowable azimuthal quantum numbers.)

The next development was the realization that proton collisional excitation may make a significant contribution. Seaton (1964) calculated proton collisional excitation of the $3s^23p$ $^2P_{\frac{3}{2}}$ level of Fe XIV. In the solar corona the mean energy of the protons is far greater than the excitation energy (2.3 eV) of the $^2P_{\frac{3}{2}}$ level, and, because for a given energy the protons have lower velocities than the electrons, the circumstances are closer to the maximum of the proton impact cross section than to the maximum of the electron impact cross section. One expects the proton reaction rate to be about 40 times that of the electrons. (This is the square root of the mass ratio, the reaction rate is $\langle v\sigma \rangle$, and $\sigma \propto v^{-2}$ for high energies.) This factor is reduced by the Coulomb field of the Fe XIV ion, which increases the electron cross section and reduces the proton cross section. Detailed calculations are needed to clarify the precise relative importance of proton and electron collisions. Seaton obtained the excitation rate for Fe XIV $3s^23p$ $^2P_{\frac{1}{2}} \rightarrow$ $^2P_{\frac{3}{2}}$. For a temperature of 10^{6}°K and 2×10^{6}°K, he found the excitation rates, $\langle v\sigma \rangle$, in Table 7-2. The importance of including cascades from higher levels and (at least for $T = 2 \times 10^{6}$°K) proton excitation is clearly demonstrated by these results.

Table 7-2. Excitation rates $\langle v\sigma \rangle$ for Fe XIV, from Seaton (1964)

	$T = 1 \times 10^6{}^\circ\text{K}$	$T = 2 \times 10^6{}^\circ\text{K}$
Electron direct excitation	1.0×10^{-9}	0.7×10^{-9}
Electron excitation via cascades	2.5×10^{-9}	3.2×10^{-9}
Proton direct excitation	0.3×10^{-9}	1.2×10^{-9}

The importance of dielectronic recombination at high temperatures was realized by Burgess (1964). In ordinary recombination we get a process such as

$$\text{Fe}^{+15}(3s) + \text{e} \rightarrow \text{Fe}^{+14}(3s, nl) + h\nu$$

In dielectronic recombination the corresponding process would be

$$\text{Fe}^{+15}(3s) + \text{e} \rightarrow \text{Fe}^{+14}(n^1l^1, nl)$$

and

$$\text{Fe}^{+14}(n^1l^1, nl) \rightarrow \text{Fe}^{+14}(3s, nl) + h\nu$$

Here (n^1l^1, nl) denotes a doubly excited state of Fe^{+14} lying in the single ionization continuum corresponding to $\text{Fe}^{+15}(3s) + \text{e}$. The (n^1l^1) makes a downward transition to any lower state, the ground state $3s$ being the important one in practice. Burgess showed that one could consider a total recombination rate to include all processes and that this rate might be 20 times the ordinary recombination rate. There is a corresponding increase in ionization rates by autoionization following an inner-shell excitation. This inverse to dielectronic recombination was studied by Bely (1967).

The importance of intermediate coupling in making radiative transitions possible between states of different total spin is well known. It was not realized until 1967 that this could be a large effect in collision cross sections. For a transition such as Fe XIII $3p^2$ $^3P_2 \rightarrow {}^1D_2$ there is a contribution arising from electron exchange. This type of contribution generally dominates processes of this kind in the low stages of ionization of an atom. For highly ionized atoms, spin-orbit interaction becomes very important and mixes states of the same total angular momentum J. In Fe XIII $3p^2$ the $^1D_2$ and $^3P_2$ states are mixed. This allows part of the collision strength of the $^3P_2$–$^3P_2$ transition (not observable—a level to itself—but having large matrix elements) to be transferred to the $^3P_2$–$^1D_2$ transition. Bely, Bely, and Vo Ky Lan (1966) obtained the collision strengths in Fe XIII as shown in Table 7-3. Exchange effects (not included in these results) are estimated to contribute about 0.03 to the collision strengths.

Table 7-3. Collision strengths in Fe XIII, from Bely, Bely, and Vo Ky Lan (1966)

	L-S coupling	Intermediate coupling
$^3P_2-^1D_2$	0.00	0.32
$^3P_0-^3P_2$	0.14	0.16

In many cases the collision cross section for the excitation of an allowed transition is proportional to the optical oscillator strength of the transition. One might expect that forbidden transitions would have small cross sections. This is by no means true in the general case. Bely and Bely (1967) computed the cross sections for the excitation of Fe XVII ions from the ground configuration $2p^6$ to the excited configurations $2p^53s$, $2p^53p$, and $2p^53d$ and to some higher states. The cross section of the transition $2p^6\ ^1S_0-2p^53p\ ^1S_0$ is large, and that of the transitions $2p^6\ ^1S_0-2p^53s\ ^3P_1$ and $^1P_1$ is smaller. The cross sections of the permitted transitions need not be the largest, at least in cases where there is a change of principal quantum number in the transition. Bely and Bely showed that when their new cross sections were used the agreement between the calculated and observed intensities of ultraviolet coronal lines was improved. We illustrate a few of the transitions of Fe XVII in Figure 7-2. There are in effect two cycles of excitation processes,

$$2p^6 \rightarrow 2p^53d \rightarrow 2p^6 \qquad \text{and} \qquad 2p^6 \rightarrow 2p^53p \rightarrow 2p^53s \rightarrow 2p^6$$

(There are many additional energy levels of Fe XVII not shown in Figure 7-2, but they do not alter this rough picture.)

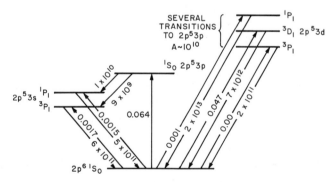

Fig. 7-2. Energy levels in Fe XVII, not to scale. Only a few levels of the $2p^53s$, $2p^53p$, and $2p^53d$ configurations are shown. The collision strengths are indicated for upward transitions and transition probabilities for downward transitions.

Another ion which has been studied in some detail is Fe XV. It is of note because, among other reasons, it is the only case for which an identified coronal line (7059 Å, in Table 7-1) arises from a transition within a configuration which is not the ground configuration. The ultraviolet resonance lines of Fe XV have been observed at 417 Å and 284 Å, these lines being surprisingly strong. Bely and Blaha (1968) showed that it is necessary to include the $3s3d$ configuration in calculations of the Fe XV line intensities. In considering the $3s^2$ $^1S \rightarrow 3s3p$ ^3P cross sections both exchange effects and intermediate coupling effects were included. The values of the collision strength are shown in Table 7-4. A further correction for resonances (mentioned below) was introduced by Bely and Blaha.

Table 7-4. Collision strength values for Fe XV, from Bely and Blaha (1968)

Transition	Exchange	Intermediate Coupling	Total
$^1S_0-^3P_0$	0.004	—	0.004
$^1S_0-^3P_1$	0.011	0.034	0.045
$^1S_0-^3P_2$	0.018	—	0.018

An important improvement in the theory of collision cross sections has arisen through the recognition of the importance of autoionizing states and resonances in electron-ion collisions. An example of this was discussed by Bely and Petrini (1966). If an electron hits a positive ion it can be captured into a doubly excited state of a once-less ionized atom, and this unstable autoionizing state can break up leaving the ion in the original stage of ionization but in an excited state. The apparent rate of excitation of this state is greater than one would expect on the basis of direct collisional excitation alone. Bely and Petrini showed that for Ca II the collision strength for $4s \rightarrow 3d$ is roughly doubled by this process in the incident energy range between the $3d$ and $4p$ levels. This phenomenon occurs (Bely and Blaha, 1968) in Fe XV for energies between the $3s3p$ ^3P and ^1P terms. Another example which has been discussed in work (as yet unpublished) by Prof. M. J. Seaton's group in London is O III, where the state $2s2p^33s^2D$ of O II lies just above the $2s^22p^2$ ^1D state of O III, and the cross section for $2s^22p^2$ $^3P \rightarrow$ ^1D in O III shows a resonance just above the threshold. There is clearly room for further investigations along these lines.

Gabriel and Jordan (1969) reported that in the solar spectrum a line at 22.09 Å is observed close to the O VII $1s^2-1s2p$ $^1P_1$ resonance line at 21.55 Å, and similar lines have also been observed near the resonance lines of C V, Ne IX, Na X, and Mg XI. The intensity is in all cases nearly as high as the resonance line. Gabriel and Jordan suggested that the lines are the $1s^2\ {}^1S_0$–$1s2s\ {}^3S_1$ transition. This transition was subsequently shown by Griem (1969) to be a magnetic dipole transition made possible in a relativistic approximation. (In nonrelativistic theory a zero magnetic dipole line strength is obtained for a transition involving a change of principal quantum number.) For O VII he estimated the transition probability of the line to be 33 $\sec^{-1}$. The $1s2s\ {}^3S_1 \rightarrow 1s^2\ {}^1S_0$ transition can also take place by a two-photon decay, but this is of lower probability: Drake, Victor, and Dalgarno (1969) obtained for this process in O VII the transition probability 0.25 $\sec^{-1}$.

The most recent development in the area of solar coronal deexcitation processes is work by Garstang (1969) on magnetic quadrupole radiation. He showed that transitions of the form $s^2\ {}^1S_0$–$sp\ {}^3P_2$ are allowed for magnetic quadrupole radiation (but for no lower order radiation). Transitions of this type occur in several ions of interest in the solar corona, and other types of magnetic quadrupole transition are

$$p^6\ {}^1S_0\text{–}p^5s\ {}^3P_2 \qquad \text{and} \qquad p^6\ {}^1S_0\text{–}p^5d\ {}^3P_2, {}^1D_2, {}^3D_2, {}^3F_2.$$

Garstang showed that in Fe IX the $3p^53d\ {}^3P_2 \rightarrow 3p^6\ {}^1S_0$ transition is an important deexcitation mechanism, and in Fe XVII the transition $2p^53s\ {}^3P_2 \rightarrow 2p^6\ {}^1S_0$ is important. The transitions $2s2p\ {}^3P_2 \rightarrow 2s^2\ {}^1S_0$ in Fe XXIII and $3s3p\ {}^3P_2 \rightarrow 3s^2\ {}^1S_0$ in Fe XV are unimportant. Lastly, in Fe XXV we have the remarkable situation that the magnetic quadrupole transition

$$1s2p\ {}^3P_2 \rightarrow 1s^2\ {}^1S_0 \qquad (A = 6.5 \times 10^9\ \sec^{-1})$$

has a higher probability than the fully allowed electric dipole transition

$$1s2p\ {}^3P_2 \rightarrow 1s2s\ {}^3S_1 \qquad (A = 5.1 \times 10^8\ \sec^{-1})$$

This magnetic quadrupole transition may well become observable in the solar spectrum in the near future: the resolving power of rocket spectrographs has now reached the point where resolution from $1s2s\ {}^3P_1 \rightarrow 1s^2\ {}^1S_0$ is possible. When this and other observations have been made it will be possible to make further checks on the theory of the solar corona and perhaps discover other mechanisms which contribute to the physical processes in the corona.

Acknowledgment

The writer's own work on magnetic quadrupole transitions was supported in part by National Aeronautics and Space Administration contract NGR-06-003-057 and by National Science Foundation Grant GP-11948.

REFERENCES

Bely, O. 1967. *Ann. Astrophys.* 30:953–57.
Bely, O., and Bely, F. 1967. *Solar Phys.* 2:285–89.
Bely, O.; Bely, F.; and Vo Ky Lan. 1966. *Ann. Astrophys.* 29:343–44.
Bely, O., and Blaha, M. 1968. *Solar Phys.* 3:563–77.
Bely, O., and Petrini, D. 1966. *Phys. Letters* 23:442–43.
Billings, D. E. 1966. *A Guide to the Solar Corona.* Academic Press, New York.
Burgess, A. 1964. *Astrophys. J.* 139:776–80.
Drake, G. W. F.; Victor, G. A.; and Dalgarno, A. 1969. *Phys. Rev.* 180:25–32.
Edlén, B. 1945. *Monthly Notices Roy. Astron. Soc.* 105:323–33.
Gabriel, A. H., and Jordan, C. 1969. *Nature* 221:947–49.
Garstang, R. H.
    1962. *Ann. Astrophys.* 25:109–17.
    1969. *Pubs. Astron. Soc. Pacific,* 81: 488–95.
Griem, H. 1969. *Astrophys. J. Letters* 156:L103–05.
Pecker, C., and Thomas, R. N. 1962. *Ann. Astrophys.* 25:100–108.
Seaton, M. J. 1964. *Monthly Notices Roy. Astron. Soc.* 127:191–94.

Negative ions of atoms and molecules were first discovered in mass spectrograph analyses. Those found included O^-, O_2^-, NO_2^-, NO_3^-, OH^-, H^-, C^-, CH^- and Li^-. Much work has been directed to understanding why negative ions exist. If E_0 is the energy of the ground state of the neutral atom and E_- is the energy of the (ground state of the) negative ion, the quantity $E_0 - E_-$ is called the *electron affinity* of the atom, and this is the energy needed to detach the extra electron from the negative ion. The negative ion is stable if $E_0 > E_-$. In addition Pauli's principle must be fulfilled, so that, for example, we would not expect helium, with a $1s^2$ closed shell, to form a negative ion with electron configuration $1s^3$. Thus we expect some limitation on the atoms which can form negative ions. In this lecture we shall review a few aspects of the study of negative ions. For a survey of the older literature, reference may be made to the book by Massey (1950), and for more recent work to the review articles by Branscomb (1957, 1962, 1964), Moiseiwitsch (1965), Ferguson (1967, 1968), Smith (1968), and Bardsley and Mandl (1968).

I should make it clear at the outset that in this chapter I shall be able to deal with only a few aspects of the subject. I have chosen to discuss primarily several lines of work in which some of my colleagues in Boulder have been engaged, and space does not permit discussion, or even mention, of many important contributions by other workers.

Electron Affinities

Efforts to determine reliable electron affinities of negative ions date back to Hylleraas' calculation on H^- in 1930. This work was refined by many later workers. Electron affinities of atoms have been determined by extrapolations along isoelectronic sequences and by photodetachment experiments. There have also been some calculations by Hartree—Fock methods with the calculations including correlation effects. Electron affinities have been determined for some molecules from lattice energies, electron impact, surface ionization, and photodetachment. An illustrative selection of results which have been obtained is given in Table 7-5. No doubly charged atomic negative ions are thought to exist. A few excited

Table 7-5. Electron affinities of selected[a] atoms and molecules

Atom	Electron Affinity (eV)	Molecule	Electron Affinity (eV)
H	0.754	H_2	0.9
He (see text)	0.08	OH	1.83
C (^{4}S)	1.25	O_2	0.15
(^{2}D)		O_3	2.9
N	not stable	C_2^-	3.1
O	1.478	SH	2.32
F	3.45	CN	3.6
Si (^{4}S)	1.39	NO	0.9
(^{2}D)	0.88 (?)	NO_2	4.0
S	2.07	NO_3	3.9
I	3.076		

[a] Selected from the compilation of Moiseiwitsch (1965), where data are given for many other atoms and molecules and references are given to the original papers, and supplemented by more recent data for He (Brehm, Gusinow and Hall, *Phys. Rev. Letters* 19:737, 1967), O (Berry, Mackie, Taylor and Lynch, *J. Chem. Phys.* 43:3067, 1965), OH (Branscomb, *Phys. Rev.* 148:11, 1966), and SH (Steiner, *J. Chem. Phys.* 49:5097, 1968). The value for H_2 is the vertical detachment energy, not the electron affinity (-3.6 eV).

states may exist, the most interesting being C^- (^{2}D) and Si^- (^{2}D). Perhaps the most interesting experimental work in this area is that of Berry, Mackie, Taylor, and Lynch (1965). They studied the emission spectrum produced by the recombination reaction $O + e \rightarrow O^- + h\nu$ in the threshold region $\lambda7800$ to $\lambda8800$ in shock-heated vapors of potassium peroxide and rubidium oxide in neon and argon carrier gas. They saw the individual thresholds $O(^3P_2) \rightarrow O^-(^2P_{\frac{1}{2}})$ at $\lambda8592$, $O(^3P_2) \rightarrow O^-(^2P_{\frac{3}{2}})$ at $\lambda8386$, and $O(^3P_1) \rightarrow O^-(^2P_{\frac{3}{2}})$ at $\lambda8276$ and deduced an electron affinity of O^- of $1.478(\pm0.002)$ eV and a doublet splitting of the $O^{-2}P$ state of $285(\pm15)$ cm^{-1}. This was the first and, so far, the only measurement of a term splitting in a negative ion.

An outstanding experiment performed by B. Brehm, M. A. Gusinow, and J. L. Hall (1967) at the Joint Institute for Laboratory Astrophysics of the National Bureau of Standards and University of Colorado in Boulder is the direct measurement of the electron affinity of the $1s2s$ ^{3}S state of helium (to form He^- $1s2s2p$ $^4P_{\frac{5}{2}}$). The existence of a metastable state of this kind had been predicted theoretically by a quantum mechanical variational calculation, but a reliable electron affinity is hard to determine because it involves the difference of two large energies.

In the experiment, positive ions of helium and deuterium are extracted from a hot-cathode arc discharge source (Fig. 7-3). The ion beam is passed through an oven containing potassium vapor at low pressure. Here some positive ions double-charge exchange to produce He⁻, H⁻, and D⁻ ions. The negative ions are separated from the positive ions by electrostatic deflection, and the negative ion beam is mass analyzed. The beam of He⁻ ions passes through an interaction chamber where it is crossed with an

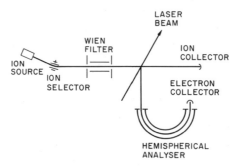

Fig. 7-3. Measurement of He electron affinity (Brehm, Gusinow, and Hall 1967). Many accelerating and decelerating lenses and electrodes are not shown.

argon ion laser operating at 4880 Å or 5145 Å. The electrons produced by photodetachment pass through a hemispherical electron analyzer and are collected. The ion current is also collected. It is thus possible to measure the kinetic energy spectrum of the photodetached electrons. An important correction is to allow for the velocity of the negative ion: The electrons must be emitted in a slightly backward direction in the center-of-mass system if they are to be measured perpendicular to the beam in the laboratory frame. Measurements on H⁻ and D⁻ ions enabled the apparatus to be calibrated. The final result was 80 ± 2 millielectron volts for the electron affinity of He $1s2s\ ^3S$, forming He⁻ $1s2s2p\ ^4P_{\frac{5}{2}}$.

Laser Two-Photon Photodetachment

Another experiment illustrating the power of lasers for negative ion experiments done at the Joint Institute for Laboratory Astrophysics by J. L. Hall, E. J. Robinson, and L. M. Branscomb (1965) was on iodine. They observed and measured the photodetachment of I⁻ by 1.785 eV photons (6943 Å). Because I⁻ has only one bound state (electron affinity 3.06 eV), two photons must be involved in the photodetachment. Measurements were made of the laser photon flux and the electrons detached, and the probability of the two-photon process was determined to be 2.7×10^{-49}

$F^2 \sec^{-1}$, where F is the flux density in photons $cm^{-2} \sec^{-1}$. In the laser used, F was about 10^{27} photons $cm^{-2} \sec^{-1}$. This result was later confirmed by theoretical calculations by Robinson and Geltman (1967), who obtained a probability $2.1 \times 10^{-49} F^2 \sec^{-1}$ by using one-electron continuum states in an assumed central field adjusted to reproduce the observed binding energies of the negative ions.

Drift Tube Experiments

We mention two experiments selected from many which have been performed with drift tubes. Chanin, Phelps, and Biondi (1962) studied the attachment of low energy (<1 eV) electrons to oxygen molecules by the reaction $e + 2O_2 \rightarrow O_2^- + O_2$. The principle of the experiment is simple (Fig. 7-4); for details the original paper should be consulted. An ultraviolet

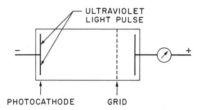

Fig. 7-4. Drift tube experiment (Chanin, Phelps, and Biondi 1962).

light pulse strikes the photocathode and releases electrons, which drift down the tube under an applied field, and some of them attach to the O_2 molecules in the gas. The O_2^- ions start to drift down the tube (much more slowly than the electrons) and their arrival at the anode is measured as a function of time (with the aid of the control grid used as a shutter). If the attachment rate is K and the electron drift velocity is w_e, we have

$$\frac{dn_e}{dt} = \frac{\partial n_e}{\partial t} + w_e \frac{\partial n_e}{\partial z} = -K[O_2]^2 n_e$$

and for a moderately long pulse $\partial n_e/\partial t \approx 0$, so that $n_e = n_e(0)\exp(-\alpha z)$ where $\alpha = K[O_2]^2/w_e$. The initial distribution of O_2^- ions will have this same $e^{-\alpha z}$ factor. The O_2^- ions drift down the tube with velocity w_i, and their rate of arrival at the anode must satisfy

$$\frac{\partial[O_2^-]}{\partial t} + w_i \frac{\partial[O_2^-]}{\partial z} = 0$$

which leads to $[O_2^-] \propto \exp(\alpha w_i t)$. Measurement of the time dependences leads to values of w_i, α, and w_e, and hence of K. The final value obtained was $K = 2.8 \times 10^{-30}$ cm^6 sec^{-1} at 300°K. Other processes (such as $e + O_2 \rightarrow O + O^- + h\nu$) are possible, but in the ionosphere of the Earth, where these processes are of interest, the three-body process dominates at all altitudes of interest.

One other example of a drift-tube experiment is that by Woo, Branscomb, and Beaty (1969) on photodetachment of electrons from O_2^-. Ions from a discharge are thermalized; they then enter the reaction region of a drift tube where they are irradiated with light of the desired spectral distribution. (In this experiment light approximating sunlight was used.) The detached electrons are collected and measured, and the ion current can also be measured, separated from the electron current by time resolution (the ion-drift velocity being much slower than the electron-drift velocity). The final result is a detachment rate for O_2^- in sunlight of 0.3 sec^{-1}.

Geometrical Hindrance in Molecular Ion Formation

One new factor which has recently come to light is the effect of geometrical factors. When electrons are mixed with N_2O or with CO_2 molecules we do not seem to get N_2O^- or CO_2^- ions. The reason seems to be connected with the fact that N_2O and CO_2 are linear molecules, whereas N_2O^- and CO_2^- are predicted to be bent (angle about 135°) from comparisons with isoelectronic molecules. The electron affinity for attachment to a linear molecule may be negative even if the electron affinity for attachment to a bent molecule is positive. This topic was discussed by Ferguson, Fehsenfeld, and Schmeltekopf (1967). It is clear that further developments can be expected in this area of molecular ion studies.

Negative Ion-Molecule Reactions

We now turn to another area of negative ion studies. Although processes involving electrons and molecules had been studied for many years, it was not until about 1967 that it became possible to measure the rates of reactions between negative ions and molecules. An outstanding series of contributions in this area has been made by a group under Dr. Eldon Ferguson at the Environmental Science Services Administration laboratories in Boulder, Colorado, U.S.A. This work included both positive and negative ion reactions with molecules (and occasionally with atoms). Excellent review articles have been written by Ferguson (1967, 1968). After some general remarks we shall discuss briefly that part of his work which deals with negative ions.

If σ is the cross section for one incident particle hitting one target particle and there are n incident particles and N target particles per unit volume with a relative velocity v between the incident and target particles, the number of collisions per second is $Nnv\sigma$. In practice there is a range of velocities (usually a Maxwell distribution), and the cross section is a function of the incident energy. We therefore write the number of collisions per second as kNn, where $k = \langle v\sigma \rangle_{av}$. This is the reaction rate for an ion-molecule reaction if every collision produces a reaction; otherwise the rate will be reduced by the probability of getting a reaction being less than 1 per collision, and this factor is included in σ when σ is calculated by quantum mechanical methods.

When an ion approaches a neutral atom or molecule they interact because of the polarization of the neutral particle by the ion. This interaction is an attraction, and to a good first approximation it may be derived from a polarization potential $-\alpha e^2/2r^4$, where α is the polarizability of the neutral particle as usually defined in quantum mechanics. It is instructive to consider the classical trajectories under such a potential. Suppose an incident charged particle has mass M and velocity v_0 at a large distance from a fixed-target neutral particle and the impact parameter is b. We write down the equations of motion in polar coordinates (r, θ) under the potential $-\alpha e^2/2r^4$ and obtain expressions for $\dot{r}$ and $\ddot{r}$. Examination of these expressions shows that there is a critical value, b_0, of the impact parameter. If $b > b_0$ the incident particle at first approaches the target, then it reaches a point where $\dot{r} = 0$ and $\ddot{r} > 0$, and finally it recedes from the target to infinity. This is elastic scattering. If $b < b_0$, then $\dot{r}$ starts negative, maintains a constant sign and a finite value which never approach zero and which behaves as r^{-2} for r tending to zero. The incident particle falls into the target. The value of b_0 is given by requiring that $\dot{r} \to 0$ and $\ddot{r} \to 0$ simultaneously at a finite distance R from the target. The classical cross section is πb_0^2. Particles striking within this area fall into the target, those outside this area are elastically scattered. Carrying out the elementary classical calculation we find that $R = b_0/\sqrt{2}$ and that

$$\sigma = \frac{2\pi}{v_0}\left(\frac{e^2\alpha}{M}\right)^{1/2}$$

Typical values of α are about $10^{-24}\,\mathrm{cm}^3$ and of M about $2 \times 10^{-23}\,\mathrm{gm}$ (reduced mass of two particles such as O and O_2). This gives a rate coefficient $k \sim 10^{-9}\,\mathrm{cm}^3\,\mathrm{sec}^{-1}$ and an impact parameter $b_0 \sim 20$ Bohr radii. Quantum mechanical effects will of course modify k somewhat, but in any event we predict cross sections of order $400\,\pi a_0^2$, which is very large, and k is therefore very large.

Many reaction rates for ion-molecule reactions have now been measured. We have space to describe only the work of Ferguson's group: Detailed references to the original papers of his group and to the work of other authors may be found in the review articles listed at the end of this chapter. The apparatus used by Ferguson is sketched in Figure 7-5. A carrier gas (usually helium) is introduced at one end, and in some experiments another gas is introduced at the side entrance 1. The gas passes through a microwave discharge which produces ions. The ionized gas enters the reaction tube, 100 cm long and 8 cm diameter, and is pumped

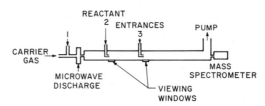

Fig. 7-5. Flowing gas system for negative ion-molecule reaction rate measurement (Fehsenfeld, Schmeltekopf, Golden, Schiff, and Ferguson 1966).

down the tube with a flow velocity about 10^4 cm per sec, the pressure being 0.3 Torr. Other reactant gases are introduced at one or both of the entrances 2 and 3 at measured flow rates small compared to that of the carrier gas. A mass spectrometer at the end of the tube samples both primary and product ions. The distance from the point of entrance of reactant to the mass spectrometer divided by the flow velocity gives the reaction time, and the flow rate of reactant gas gives the density of reactant gas molecules. The velocity in the tube was determined by following the progress of a pulse of plasma down the tube. Under typical conditions electron and ion densities of 5×10^{10} cm^{-3} are obtained in the tube.

We illustrate the reduction of the experimental data by considering a simple positive ion experiment. Helium is the carrier gas, and O_2 is introduced at entrance 2. The microwave discharge in the helium produces He$^+$ ions and He (2^1S) and He (2^3S) metastable atoms. We consider the reaction

$$\text{He}^+ + O_2 \rightarrow \text{He} + O + O^+ \qquad (7\text{-}1)$$

In a steady flow the He$^+$ concentration (denoted by [ ]) satisfies the differential equation

$$v\frac{\partial}{\partial z}[\text{He}^+] = -A[\text{He}^+] - k[\text{He}^+][O_2]$$

where v is the flow velocity, z the distance along the tube, and A is a constant which allows for diffusive loss of He$^+$ to the walls. If we introduce

O_2 so that $[O_2] \gg [He^+]$ we may treat $[O_2]$ as constant along the flow tube, and then the solution of the above equation is

$$\log \{[He^+]_0/[He^+]_\tau\} = \{A + k[O_2]\}\tau$$

where the reaction time $\tau = l/v$ for a distance l from entrance 2 to the mass spectrometer, and the subscripts 0 and τ denote the initial and final He^+ concentrations. We measure $[He^+]$, recorded by the mass spectrometer, as a function of $[O_2]$, determined by the O_2 gas flow. A plot of $\log [He^+]_\tau$ against $[O_2]$ gives k. For the $He^+ + O_2$ reaction, Ferguson's group found $k = 1.5 \times 10^{-9}$ cm^3 sec^{-1}.

There are many additional complications in this work, and much further discussion and experimental details may be found in Fehsenfeld et al. (1966) and in Golden et al. (1966). The most important general conclusion which emerges from this work is the large size of k which is obtained for many reactions. In terms of the elementary classical concepts discussed above every collision within the cross section πb_0^2 must have a probability of order 1 of giving a reaction. This is contrary to what might be expected on the basis of the Massey adiabatic hypothesis (McDaniel 1964, p. 240), that when two atomic systems approach with a velocity low compared with typical electron orbital velocities the electrons in the atoms will adiabatically adjust to the perturbation and the probability of a reaction (electronic transition) will be small. This hypothesis is obeyed for atomic nonresonant charge transfer processes (such as $He^+ + Ar \rightarrow He + Ar^+$), for which the cross sections are small at low energies. The reasons for the large cross sections found in the ion-molecule work are not fully understood, but the presence of large numbers of vibrational and rotational energy levels in a molecule may increase the probability of near resonances, and in some cases the formation of intermediate complexes may be responsible for the large rate.

Before leaving the general discussion of Ferguson's work we mention that in the discharge there is also produced a significant concentration of metastable $He (2^3S)$. If we introduce O_2 at entrance 2 we get a reaction of the Penning type

$$He (2^3S) + O_2 \rightarrow He + O_2^+ + e$$

We can then work with the O_2^+ ions and introduce another reactant at entrance 3. Another possibility is to introduce argon in entrance 1 and O_2 at entrance 2, when we get $Ar^+ + O_2 \rightarrow Ar + O_2^+$. This method has the advantage of producing O_2^+ ions without contaminating O^+ ions (which cannot be produced for energy reasons in this reaction, but which would be produced by Reaction 7-1 if helium was used without argon). One other possible extension is to pass a reactant gas through a subsidiary microwave discharge before entering at entrance 2. In this way it is possible to introduce atomic O and N after dissociation from O_2 and N_2 in the subsidiary discharge.

We now mention the applications of this method to negative ion reactions. The only modifications are in the production of the ions. If O_2 is added through entrance 1, the resulting plasma contains O^- ions produced by the dissociative attachment reaction

$$e + O_2 \to O^- + O$$

If N_2O is introduced through entrance 1, we get NO^- ions. Then introducing O_2 through entrance 2 gives O_2^- ions by the reaction

$$NO^- + O_2 \to O_2^- + NO$$

The introduction of H_2O through entrance 1 gives OH^- ions. In an experiment on H^-, H_2 was introduced through entrance 1, the discharge producing H atoms, and the H^- ions were produced by the reaction

$$NH_3 + e \to H^- + NH_2$$

using electrons from a hot filament.

Some negative ion reaction rates are given in Table 7-6. So far experiment in this area has far outpaced theory. Ferguson (1968) discusses some attempts to improve on the simple classical model, but so far there do not seem to have been any attempts to explain negative ion-molecule reaction rates by detailed mechanisms or models.

Negative Ions in the Ionosphere

The Earth's upper atmosphere contains many positive ions and some negative ions. Above 100 km, positive ions predominate (species include O^+, N^+, NO^+, O_2^+, and N_2^+ with H^+ and He^+ above 500 km). Below 100 km, down to about 60 km, the ionospheric D region is believed to contain many negative ions. The densities at 75 km (particles per cm^3) are roughly (rounded to nearest power of 10) total 10^{15} (these are mostly O_2 and N_2 in a ratio 1:4) and lesser constituents: O, 10^{11}; CO_2, 10^{11}; O_3, 10^9; H_2, 10^9; H_2O, 10^9; NO, 10^8; H, 10^7; NO^+, 10^3; electrons, 10^3; O_2^+, 10^2; O^-, 10; O_2^-, 1; O_3^-, 10^{-2}. The negative ions are thought to be far more important for ionospheric physics (radio propagation and absorption) than their low densities might suggest. The study of these negative ions is only just beginning. Up to 1967 there were very few reaction rates known and no direct measurements by rocket techniques of the D-region negative-ion composition.

Table 7-6. Negative ion reaction rates[a]

Type	Reaction	Rate[b]
Charge transfer	$O^- + O_3 \rightarrow O_3^- + O$	5.0×10^{-10}
	$O_2^- + O_3 \rightarrow O_3^- + O_2$	3.0×10^{-10}
	$O^- + NO_2 \rightarrow NO_2^- + O$	1.2×10^{-9}
	$O_2^- + NO_2 \rightarrow NO_2^- + O_2$	8.0×10^{-10}
Associative Detachment	$O^- + O \rightarrow O_2 + e$	2.0×10^{-10}
	$O^- + NO \rightarrow NO_2 + e$	1.8×10^{-10}
	$O_2^- + O \rightarrow O_3 + e$	3.0×10^{-10}
	$H^- + H \rightarrow H_2 + e$	1.3×10^{-9}
Ion-Atom Interchange	$O_3^- + NO \rightarrow NO_3^- + O$	1.0×10^{-11}
	$O_3^- + CO_2 \rightarrow CO_3^- + O_2$	4.0×10^{-10}
	$CO_3^- + O \rightarrow O_2^- + CO_2$	8.0×10^{-11}
	$CO_3^- + NO \rightarrow NO_2^- + CO_2$	9.0×10^{-12}
Three-Body	$O^- + 2O_2 \rightarrow O_3^- + O_2$	$4 \times 10^{-31} \, cm^6 \, sec^{-1}$
Attachment	$e + 2O_2 \rightarrow O_2^- + O_2$	$1.4 \times 10^{-30} \, cm^6 \, sec^{-1}$
	$e + O_3 \rightarrow O^- + O_2$	4.0×10^{-11}
Photodetachment (sunlight)	$hv + O^- \rightarrow O + e$	$1.4 \, sec^{-1}$
	$hv + O_2^- \rightarrow O_2 + e$	$0.4 \, sec^{-1}$

[a] Selected from Ferguson (1967), where five additional negative ion reaction rates are given. Results for many other reactions are given in Ferguson (1968), Fehsenfeld et al. (1967), and LeLevier and Branscomb (1968).

[b] In $cm^3 \, sec^{-1}$ unless otherwise indicated.

The main D-region negative-ion production mechanism is believed to be

$$e + 2O_2 \rightarrow O_2^- + O_2$$

which we discussed earlier. The rate of this reaction depends on the square of the O_2 density, which explains why it is important only at lower altitudes. There are some other reactions which may be important in the D region, and a selection of these are listed in Table 7-6. The process

$$e + O_3 \rightarrow O^- + O_2$$

may be a significant source of O^- ions.

A theory of the negative-ion chemistry requires us to identify the loss mechanism for O_2^- and O^- ions. Two likely mechanisms are

$$O_2^- + O \rightarrow O_3 + e \quad \text{and} \quad O^- + O \rightarrow O_2 + e$$

A development of much interest is the possibility of producing other negative ions by reactions such as

$$O^- + O_3 \rightarrow O_3^- + O \qquad\qquad O_3^- + NO \rightarrow NO_3^- + O$$

$$O^- + NO_2 \rightarrow NO_2^- + O \qquad\qquad O_3^- + CO_2 \rightarrow CO_3^- + O_2$$

and there are others. A suggested scheme interlinking the ions is shown in Figure 7-6. This was discussed in Ferguson (1967) and in LeLevier and Branscomb (1968). The loss processes for the NO_2^-, NO_3^-, and CO_3^- are not yet known, but these ions are comparatively stable, and they may

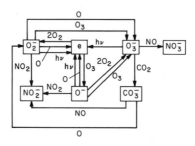

Fig. 7-6. Possible negative-ion reactions in the ionospheric D region (Ferguson 1967; LeLevier and Branscomb 1968).

exist in substantial abundance in the D region. The greatest need in this area is some rocket measurements of negative-ion abundances in the D region. There are many associated problems, such as variability of the composition of the D region with the solar cycle, diurnal variations, and latitude effects. The large abundance of neutral particles in the D region also increases the difficulty of measuring negative ion concentrations.

Negative Ions in Astrophysics

It has been known for many years that photodetachment from the H^- ion is an important contributor to the opacity of the cooler stellar atmospheres. One problem of interest is how the H^- : H ratio is maintained. Pagel suggested that the reaction

$$H^- + H \rightarrow H_2 + e$$

is important. The rate of this reaction has been measured (Schmeltekopf, Fehsenfeld, and Ferguson 1967) and found to be $k = 1.3 \times 10^{-9}$ cm^3 sec^{-1}. This is a very fast rate. There are other collisional processes involving H^-, for example $H^- + e \rightarrow H + 2e$, but these are believed to be unimportant in stellar atmospheres. The cross section for $H^- + e \rightarrow H + 2e$ was measured by Tisone and Branscomb (1968) in a crossed beam experiment.

They found a maximum cross section of $50\pi a_0^2$ at 20 eV energy. The cross section is probably several orders of magnitude smaller at thermal energies. The H^- + H reaction is rapid compared with photoionization processes (H^- + $h\nu \to$ H + e), and the concentration of H^- appears to be collisionally controlled and hence in local thermodynamical equilibrium.

Other molecular ions (OH^-, Cl^-, for example) are predicted to occur in late-type stellar atmospheres, on the basis of equilibrium calculations, but they do not yet appear to have been directly observed.

Acknowledgment

The many years of work of my colleagues in the various laboratories in Boulder has been made possible by generous support from many sources, including the Advanced Research Projects Agency, the Defense Atomic Support Agency, the National Bureau of Standards, the Environmental Science Services Administration, and the National Bureau of Standards. My own work in preparing this lecture was supported in part by National Science Foundation Grant GP-11948.

SELECTED REFERENCES—BOOKS AND REVIEWS

Bardsley, J. N., and Mandl, F. 1968. *Reports on Progress in Physics*, 31(2):471–531.
Branscomb, L. M.
    1957. Negative Ions, *Advances in Electronics and Electron Physics*, ed. L. Marton, 9:43–94. New York: Academic Press.
    1962. Photodetachment, *Atomic and Molecular Processes*, ed. D. R. Bates, pp. 100–40. New York: Academic Press.
    1964. A review of Photodetachment and Related Negative Ion Processes Relevant to Aeronomy, *Ann. Geophys.* 20:88–104.
Ferguson, E. E.
    1967. Ionospheric Ion-Molecule Reaction Rates, *Reviews of Geophysics* 5:305–27.
    1968. Thermal Energy Ion-Molecule Reactions, *Advances in Electronics and Electron Physics*, ed. L. Marton 24:1–50. New York: Academic Press.
Massey, H. S. W. 1950. *Negative Ions*. London, U.K.: Cambridge University Press.
McDaniel E. W. 1964. *Collision Processes in Ionized Gases*. New York: Wiley.
Moiseiwitsch, B. L. 1965. Electron Affinities of Atoms and Molecules, *Advances in Atomic and Molecular Physics* 1:61–83. New York: Academic Press.
Smith, S. J. 1968. Photodetachment, *Methods of Experimental Physics*, ed. L. Marton, *Atomic and Electron Physics*, Vol. 7, Part A, ed. B. Bederson and W. L. Fite, pp. 179–208. New York: Academic Press.

SELECTED REFERENCES—INDIVIDUAL ARTICLES

Multiplet Splitting
Berry, R. S.; Mackie, J. C.; Taylor, R. L.; and Lynch, R. 1965. *J. Chem. Phys.* 43:3067–74.

Laser Photodetachment
Brehm, B.; Gusinow, M. A.; and Hall, J. L. 1967. *Phys. Rev. Letters* 19:737–41.
Robinson, E. J., and Geltman, S. 1967. *Phys. Rev.* 153:4–8.

Two-Photon Photodetachment
Hall, J. L.; Robinson, E. J.; and Branscomb, L. M. 1965. *Phys. Rev. Letters* 14:1013–16.

Drift-Tube Experiments

Chanin, L. M.; Phelps, A. V.; and Biondi, M. A. 1962. *Phys. Rev.* 128:219–30.
Woo, S. B.; Branscomb, L. M.; and Beaty, E. C. 1969. *J. Geophys. Res.* 74:2933–40.

Geometrical Hindrance

Ferguson, E. E.; Fehsenfeld, F. C.; and Schmeltekopf, A. L. 1967. *J. Chem. Phys.* 47:3085–86.

Ion-Molecule Reactions

Fehsenfeld, F. C.; Ferguson, E. E.; and Schmeltekopf, A. L. 1966. *J. Chem. Phys.* 45:1844–45.
Fehsenfeld, F. C.; Schmeltekopf, A. L.; Golden, P. L.; Schiff, H. I.; and Ferguson, E. E. 1966. *J. Chem. Phys.* 44:4087–94.
Fehsenfeld, F. C.; Schmeltekopf, A. L.; Schiff, H. I.; and Ferguson, E. E. 1967. *Planet. Space Sci.* 15:373–79.
Golden, P. D.; Schmeltekopf, A. L.; Fehsenfeld, F. C.; Schiff, H. I.; and Ferguson, E. E. 1966. *J. Chem. Phys.* 44:4095–103.
Schmeltekopf, A. L.; Fehsenfeld, F. C.; and Ferguson, E. E. 1967. *Astrophys. J.* 148:L155–L156.

Electron Impact Detachment

Tisone, G. C., and Branscomb, L. M. 1968. *Phys. Rev.* 170:169–83.

Ionospheric Negative Ions

LeLevier, R. E., and Branscomb, L. M. 1968. *J. Geophys. Res.* 73:27–41.

3. FORBIDDEN ATOMIC TRANSITIONS

Many years ago atomic spectral transitions were discovered which violated the usual selection rules, and such transitions became known as forbidden transitions. The first transitions to be recognized as such were the $^2D-^2S$ transitions in the alkali metal spectra, the

$$6\,^3P_2-6\,^1S_0 \qquad \text{and} \qquad 6\,^3P_0-6\,^1S_0$$

transitions in HgI and the auroral line $^1S_0-^1D_2$ in O I. The discovery by Bowen in 1928 that many of the strongest lines in gaseous nebulae (lines of ions such as O II, O III, N II, S II, and Ne III, to name but a few) are forbidden transitions gave a great impetus to the subject. Other forbidden transitions were found in peculiar stars. In 1942 Edlén identified the previously unidentified emission lines in the visible spectrum of the solar corona as being caused by forbidden transitions in highly ionized atoms (such as Fe X, Fe XI, Fe XIV, and Ni XII).

Calculations of Transition Probabilities

Many of the important forbidden transitions mentioned above are due to magnetic dipole radiation, and others are due to electric quadrupole radiation. These types of radiation have been studied in detail: reviews by Garstang (1962, 1969) may be consulted for the original references. [For more recent work see also the next chapter in this book.] Transition probabilities have been calculated for most of the astrophysically interesting lines and for most lines which can be produced in the laboratory. It is very difficult to assess the accuracy of forbidden-transition probabilities. The evidence from the experimental work discussed below is encouraging. There is room for further studies of the effects of configuration interaction and of electron correlation on forbidden line strengths. An elaborate study of some cases has just been carried out with the theory of electron correlation by Sinanoğlu. For these and additional considerations, the reader is referred also to the next chapter.

The Fe II Problem

There has been considerable debate as to the correct amount that measures the abundance of iron in the solar atmosphere. Determinations from the study of ordinary Fe I lines in the solar photosphere seem to indicate an abundance of iron about seven times lower than the abundance given by ultraviolet lines from the solar corona. One possible check is to use the faint forbidden lines of Fe II which are observable in the solar photosphere. A great deal of effort has been devoted to observing these lines, and their identification seems beyond question. There are uncertainties in the model atmospheres used in the interpretation of the lines, but it is hard to see how these could lead to such a large error. It is equally hard to see how the transition probabilities of the [Fe II] lines could be in error by an order of magnitude. Thackeray made some extensive observations of the emission line star η Carinae, which shows a rich spectrum of Fe II, [Fe II], and other ions. He compared his intensities of [Fe II] and [Ni II] lines with the theoretical transition probabilities. The agreement is very good, and on the basis of the theoretically predicted intensities many new identifications could be made. The excellent agreement between theory and observation leads us to think that there are no gross errors in the relative intensities of lines within the various multiplets. One can conclude little from the comparison about the absolute intensities, except to say that there seem to be no major discrepancies. However, the magnetic dipole line strengths do not depend on the radial wavefunctions but are largely determined by spin-orbit interaction, and this seems reasonably well known in Fe II from energy level calculations. Thus we do not think that there can be gross errors in the magnetic dipole transition probabilities. The electric quadrupole transition probabilities depend on the radial quadrupole integrals (departures from L-S coupling are usually minor in many of the transitions in which electric quadrupole radiation predominates), and some uncertainty exists in these integrals but hardly enough to explain more than a factor 2 at worst, and probably much less. The evidence presently available suggests that something is wrong with the low value for the iron abundance obtained from the solar photosphere, but this remains an unsolved problem.

Experimental Intensity Measurements

There have been several laboratory investigations which lead to a comparison of theoretical and experimental intensities. When both magnetic dipole and electric quadrupole radiation are present in a particular spectrum line the corresponding intensities of the Zeeman components are not additive (for observations in a particular direction), but are modified by interference effects which vary with the direction of observation. Hults in 1966 made a careful study of the $^1D_2-^3P_1$ line in

[Pb I] and the $^2P_{\frac{3}{2}}-^2P_{\frac{1}{2}}$ line in [Pb II]. From observations of the intensities of Zeeman components it was possible to determine the percentage contribution to each of the whole lines made by electric quadrupole radiation. Hults obtained 4% and 3% for the two lines, compared with 3.2% and 5.0%, respectively, from theory. This indicates that there is no major discrepancy between theory and experiment and is a general confirmation of the theory.

In another experiment, Husain and Wiesenfeld (1967) studied the flash photolysis of trifluoroiodomethane, which produces a large concentration of iodine atoms in the upper $^2P_{\frac{1}{2}}$ level of the ground doublet term. This decays to the lower $^2P_{\frac{3}{2}}$ level by a magnetic dipole transition at 1.315 microns wavelength. After correction had been made for collisional deactivation effects and diffusion effects there was found a measurable residual decay rate, which was attributed to the spontaneous radiative decay rate of the $^2P_{\frac{1}{2}}$ level. The result was 22 (± 6) sec^{-1} for the spontaneous transition probability, which may be compared with the theoretical value of 8 sec^{-1}. This discrepancy should not be regarded as serious in view of the difficulties and pioneering nature of the experiment. There is clearly room for more work of this type.

Several investigations have checked the relative intensities of pairs of lines arising in emission from a common upper level. In the [O I] spectrum the lines $\lambda 5577$ ($^1S_0-^1D_2$) and $\lambda 2972$ ($^1S_0-^3P_1$) originate from the same upper $^1S_0$ level, so that their transition probabilities are in the same ratio as their photon emission rates. The corresponding pair of lines in the [S I] spectrum has been measured. In [O I] the ratio of the intensities of the lines $\lambda 6300$ ($^1D_2-^3P_1$) and $\lambda 6363$ ($^1D_2-^3P_2$) from the common upper level $^1D_2$ has also been measured. In all these cases the agreement between theory and experiment is excellent.

Magnetic Quadrupole Radiation

In 1964 Mizushima drew attention to the possible occurrence of magnetic quadrupole (and higher magnetic multipole) radiation for $\Delta S = \pm 1$, which is forbidden for electric multipole radiation and for magnetic dipole radiation in L-S coupling, but which is allowed for magnetic quadrupole radiation provided that there is in addition a parity change and that $\Delta J = 0, \pm 1, \pm 2$ ($0 \leftrightarrow 0, 0 \leftrightarrow 1, \frac{1}{2} \leftrightarrow \frac{1}{2}$), and $\Delta L = 0, \pm 1$ ($0 \leftrightarrow 0$). In many cases where magnetic quadrupole radiation occurs electric dipole radiation made possible by spin-orbit interaction also occurs and is stronger, effectively preventing the observation of the magnetic quadrupole radiation. This does not happen for $\Delta J = \pm 2$, and of the transitions satisfying this the transition $s^2\ ^1S_0-sp\ ^3P_2$ is perhaps the best case for a search for magnetic quadrupole radiation. A competition with nuclear-spin-induced radiation is possible in this case, and it is necessary to calculate both to determine which is dominant. It turns out that for Mg I ($3s^2\ {}^1S_0$–$3s3p\ {}^3P_2$) and for Zn I ($4s^2\ {}^1S_0$–$4s4p\ {}^3P_2$) magnetic quadrupole radiation predominates. For similar transitions in Cd I and Hg I nuclear-spin-induced radiation is dominant. The Zn I line has been observed in the laboratory, the only recorded observation of a pre-dominantly magnetic quadrupole line. The Mg I line has been identified in the planetary nebula NGC 7027. Magnetic quadrupole radiation may be important for some highly ionized atoms, and this possibility is mentioned in this volume in my chapter on excitation processes in the solar corona.

In conclusion mention should be made of two other types of forbidden transition. One involves two-quantum processes. The transition $2s\ {}^2S \rightarrow 1s\ {}^2S_{\frac{1}{2}}$ in hydrogen and hydrogenlike ions and the transition $2\ {}^1S_0 \rightarrow 1\ {}^1S_0$ in He I and heliumlike ions both occur by the spontaneous emission of two photons. The other type involves magnetic dipole radiation in a relativistic approximation. This transition, $2\ {}^3S_1$–$1\ {}^1S_0$ in helium and heliumlike ions, has been observed in the solar corona in the ion O VII, and is mentioned in my solar corona chapter in this volume. Two photon processes can in principle compete with the relativistic magnetic dipole process but are found on investigation to be extremely weak, and quite negligible for O VII.

REFERENCES

These are review articles and contain extensive bibliographies of original papers.
Garstang, R. H.
    1962. in *Atomic and Molecular Processes*, ed. D. R. Bates. New York: Academic Press, chapter 1.
    1969. *Mém. Soc. Roy. Sci. Liège* Sér. V. 17:35.

8

C. NICOLAIDES AND O. SİNANOĞLU

Atomic Transition Probabilities
New Experimental and Theoretical
Results and Their Comparison

The measurement or the calculation of atomic transition probabilities for many electron systems has been a difficult problem in atomic physics. A search through the literature can easily reveal a variety of results that differ by orders of magnitude and that have been obtained by a variety of methods. Recently, however, new approaches both in theory and experiment have been developed. These now provide, at least for the first-row atoms so far, consistent and very accurate results on absolute transition probabilities with an often better than 10% to 15% error. This unprecedented agreement of theory with experiment is very encouraging since it provides a strong indication of the correctness of both theoretical and experimental methods, which can now be used to predict safely various spectral-line transition probabilities.

In the discussion to follow, we shall first give, for the sake of completeness, the basic formulas from the theory of allowed transition probabilities. We then briefly describe the experimental methods used to measure transition probabilities and outline the new theoretical method (based on the theory of atomic structure including electron correlation by Sinanoğlu) which gives accurate transition probabilities in agreement—within experimental error—with recent experimental results. Our main interest will be the allowed (electric dipole) transitions while a brief account of new results on forbidden electric quadrupole transitions will also be given.

ALLOWED TRANSITIONS

Basic Formulas

For first row atoms, L-S coupling is assumed. The allowed transition probability from a lower state (characterized by γLSM_LM_S, γ denoting the configuration, to an upper state, f, with quantum numbers, $\gamma'L'S'M'_LM'_S$)

139

is given in atomic units ($e = \hbar = m = 1, c = 137$)

$$A_{if} = \frac{4}{3} \frac{(E_f - E_i)^3}{c^3} \frac{1}{g} S(i, f) \tag{8-1}$$

where $g = (2L + 1)(2S + 1)$ is the weight (degeneracy) of the initial state and $S(i, f) = S(f, i)$ is the multiplet line strength, a quantity solely dependent on the matrix element of the dipole operator and given by:

$$S(i, f) = \sum_{M_L = -L}^{L} \sum_{M_S = -S}^{S} \sum_{M'_L = -L'}^{L'} \sum_{M'_S = -S'}^{S'} \left| \frac{\int dx_1 \cdots dx_N \psi_i^* \mathbf{R} \psi_f}{(\langle \psi_i | \psi_i \rangle \langle \psi_f | \psi_f \rangle)^{\frac{1}{2}}} \right|^2 \tag{8-2}$$

where $\mathbf{R} = \sum_{i=1}^{N} \mathbf{r}_i$ is the electric dipole length operator and N is the number of electrons. The quantity most frequently used is the absorption oscillator strength defined by

$$f_{if} = \frac{2}{3} \frac{(E_f - E_i)}{(2L + 1)(2S + 1)} S(i, f) \tag{8-3}$$

In the case where the upper state can decay to more than one lower state, the total transition probability is just the sum of the individual transition probabilities. There exists then a relationship between the total transition probability and the life time of the upper state as shown by the Einstein theory of spontaneous emission:

$$A_{\text{total}} = \sum_i A_{if} = \frac{1}{\tau_f} \tag{8-4}$$

This is the relationship used to obtain transition probabilities from lifetime measurements of excited states. Obviously in the case where there is only one lower state the sum simplifies to a single term, expressing the transition probability of a single spectral line. Typical values for transition probabilities of allowed transitions are of the order of 10^7–10^9 sec^{-1}.

The transition probability of a spectral line is also related to the total intensity I_{if} of that line. The quantity I_{if} expresses the energy emitted per second per square centimeter per steradian from a thin layer of length L and is given by

$$I_{if} = \frac{L}{4\pi} A_{if} N_i h\nu \tag{8-5}$$

where N_i is the number population of the excited state. This relationship is basic to all absorption and emission experiments where measurement of the intensity absorbed or emitted by the species of interest is the main objective. We shall return to this point in our discussion of experimental methods.

A word should be said about the matrix element $\langle \psi_i | \mathbf{R} | \psi_f \rangle$ which appears in the multiplet line strength. Using the Heisenberg equation of motion for a dynamical variable:

$$\frac{dA}{dt} = \frac{i}{\hbar}[H, A] \tag{8-6}$$

and the "off-diagonal hypervirial theorem," which is valid for an arbitrary Hermitian operator A when exact eigenfunctions are used:

$$\langle \psi_f | [H, A] | \psi_i \rangle = (E_f - E_i) \langle \psi_f | A | \psi_i \rangle \tag{8-7}$$

we can obtain alternative forms to the dipole length operator expression. These are the dipole velocity operator

$$\mathbf{V} \equiv \sum_{i=1}^{N} \mathbf{V}_i$$

and dipole acceleration operator $A' = AZ$, where

$$\mathbf{A} \equiv \sum_{i=1}^{N} \frac{\vec{r}_i}{r_i^3}$$

and it is easy to show that they are related by the equations

$$\langle \psi_f | \mathbf{V} | \psi_i \rangle = -\frac{m}{\hbar^2}(E_f - E_i)\langle \psi_f | \mathbf{R} | \psi_i \rangle = -Ze^2(E_f - E_i)^{-1}\langle \psi_f | \mathbf{A} | \psi_i \rangle \tag{8-8}$$

The minus sign reflects the anti-Hermitian nature of the dipole velocity operator. The corresponding expressions for the oscillator strengths are then in atomic units:

$$f_r = \frac{2}{3} \frac{(E_f - E_i)}{(2L + 1)(2S + 1)} |\langle \psi_f | \mathbf{R} | \psi_i \rangle|^2 \tag{8-9}$$

$$f_v = \frac{2}{3} \frac{(E_f - E_i)^{-1}}{(2L + 1)(2S + 1)} |\langle \psi_f | \mathbf{V} | \psi_i \rangle|^2 \tag{8-10}$$

$$f_a = \frac{2}{3} Z^2 \frac{(E_f - E_i)^{-3}}{(2L + 1)(2S + 1)} |\langle \psi_f | \mathbf{A} | \psi_i \rangle|^2 \tag{8-11}$$

where the sums over the degenerate levels are omitted. When exact wavefunctions of the nonrelativistic Hamiltonian are used, the above three expressions must give identical results. Such agreement is obtained for the hydrogen and helium atoms for which the exact and nearly exact nonrelativistic wavefunctions are known. For larger atoms it is believed that a close agreement among the results of the three expressions would imply very accurate many-electron wavefunctions. However the results in the literature show large discrepancies with the three operators. These operators "weigh" the configuration space at different distances from the nucleus. The length operator weighs the outer region more heavily, the velocity operator the region at intermediate distances, and the acceleration operator the inner region (Chandrasekhar 1945). The wavefunctions determined variationally are probably most accurate in the intermediate region. For the transitions under consideration, like those from

$$1s^2 2s^n 2p^m \rightarrow 1s^2 2s^{n-1} 2p^{m+1}$$

which involve electrons in the intermediate and outer regions, it is customary to use the dipole length and dipole velocity forms. These occasionally agree with each other. However the result need not be the "correct" one, e.g., agreement with an accurately determined experimental value. There are examples where the calculated oscillator strengths are off by about a factor of 2 from the experimentally determined ones in spite of the excellent agreement between the two. (See, for example, the inaccurate RHF, f_r, f_v values for the C II ^{2}P–^{2}D transition in Table 8-1. Such occasional agreement should therefore be considered necessary but not sufficient. Results in the earlier literature obtained with the acceleration form are consistently larger by at least an order of magnitude and are unreliable. It is customary to use the f_r and f_v instead. (We have recently obtained good results also with the f_a, which, however, will not be discussed here.)

In Equations 8-9 and 8-10, f_r and f_v depend on the excitation energy which, however, cancels out if the geometric mean $f_{\overline{rv}}$ is used.

$$f_{\overline{rv}} = \sqrt{f_r f_v} = \frac{2}{3} \frac{1}{(2L+1)(2S+1)} |\langle \psi_f | \mathbf{R} | \psi_i \rangle| |\langle \psi_f | \mathbf{V} | \psi_i \rangle| \qquad (8\text{-}12)$$

This form then is often more advantageous to use.

With a sufficiently accurate wavefunction, as in the Sinanoğlu–Westhaus work, the f_r and f_v agree closely with each other as well as with experiment. In this case, of course, either form or the $f_{\overline{rv}}$ may be used.

The ultimate and very sensitive test of an N-electron wavefunction is provided by the acceleration form f_a but is not essential for obtaining an $f_{\overline{rv}}$ accurate to within about 10%. The f_a form has just been studied by C. Nicolaides with quite encouraging results. But this aspect of the theoretical results will be discussed and reported elsewhere after a few more calculations have been completed.

Experimental Methods

The establishment of experimentally determined accurate multiplet oscillator strengths has always been a difficult task. Especially for transitions in first-row atoms and ions which are found in the ultraviolet spectrum, traditional methods have not been applicable or capable of producing results of the desired accuracy. Recently, however, improvement and application of the "emission" and "phase-shift" methods, and especially the invention of "beam-foil" spectroscopy, have produced a great number of reliable transition probability results, especially for the first-row atoms and their ions which, until a short time ago, had been uninvestigated. We shall briefly discuss these experimental methods, which have provided most of the existing results. Excellent reviews of the details of the experiments have been given by Foster (1964) and Wiese (1968) and by Bashkin and others in the two volumes entitled *Beam-Foil Spectroscopy* edited by Bashkin (1968).

One can divide the experimental methods into two groups: (a) those that involve the measurement of the intensity of the radiation of a particular spectral line and (b) those which measure the life time of an excited state directly. The first category is vapor density dependent, which alone introduces a sizable uncertainty in the final result. Methods in the second group are density independent. There are of course other factors introducing uncertainties in experiments in both groups. However, the life time experiments have been unprecedented in producing accurate values for transitions in the first-row atoms, the uncertainty considered to be usually within 15%. As is implied by Equation 8-5, the transition probability, which is the quantity characteristic of a particular line, is a function of the experimentally measured quantities, the intensity of radiation, the vapor density of the species under consideration, and the temperature of the system which is necessary for the evaluation of the Boltzmann factor used to statistically determine the number population of the excited state. In a typical absorption experiment one measures the reduction of the intensity of the radiation generated in a continuum source as it passes through a thin layer of a known number of atoms. In the case of excited states, the thin layer of atoms is heated to sufficiently populate the excited state. In this case, however, Boltzmann factors are needed, and also the problem of stimulated emission becomes important. Technical problems with available equipment limit the allowable temperatures to a maximum of 3000°K (Wiese 1968). In general, uncertainties in the vapor density together with other difficulties render it a very difficult task to make absorption measurements for transitions in the first row.

Another density-dependent experiment, used extensively in Russia, is the Hook method (Foster 1964). It is based on the theory of anomalous dispersion, which predicts a sharp change in the refractive index of a gas at regions of resonance. This method is in theory more useful than absorption experiments because it is applicable to excited as well as to ground states. However, no accurate measurements on first-row atomic transitions have been performed thus far.

A third method, which is widely used and which, although density and temperature dependent, is very promising, is the emission-radiation measurements from different kinds of shock tubes and thermal arcs. The wide range of temperature achieved (5,000°K to 50,000°K) together with the high electron densities that these sources can produce (Wiese 1968) enable the experimentalist to create thermally excited atomic and even multiply ionized lines and perform measurements of oscillator strengths with a possible error of about 15% to 20% under extremely favorable conditions for many elements of the periodic table. Errors in plasma diagnostics and intensity measurements contribute to the uncertainty of the final values (Wiese 1968).

The second group of experiments, which measure life times of excited states, are by far the most accurate and widely applicable methods. The phase-shift method and especially the beam-foil technique have produced an abundance of accurate results for first-row atoms and their ions. In principle, their applicability extends to all elements.

In the phase-shift method one excites the low-density gas to be studied by a continuous sinusoidally modulated light source or by an electron beam. The scattered radiation is also modulated, and the phase shift between exciting and fluorescent radiation is measured. The phase shift δ, thus measured, is related to the life time of the excited state by the simple relation $\tan \delta = w$ where w is the modulation frequency. Optical excitation has the advantage that it eliminates repopulation of the excited level through cascade effects while electronic excitation has the advantage that it is not restricted to resonance line measurements (Foster 1964).

A new method, developed by Bashkin (1968) and others, has been very successful in measuring accurate absolute oscillator strengths for neutral and multiply ionized lines in the first row. This is the method of beam-foil spectroscopy. Its theoretical principle is simple: A high velocity mono-energetic beam of ions is allowed to impinge on a very thin (~ 1000 Å) carbon or beryllium foil. The emerging beam contains ions of different degrees of ionization, many of them being in excited states. The charge states of these ions are determined by having an electric field, vertical to the beam, which separates beam components of different net charge (Bashkin et al. 1964). The excited states decay by emitting radiation, corresponding to a particular transition, which is recorded along the direction of the beam over the distance which is traveled by the ions away from the foil. The number of protons n_k emitted when a particular transition occurs (Bickel 1968) is given by:

$$n_k = n_i A_{ij} = n_{i0} A_{ij} e^{-t/\tau_i}$$

$$= n_{i0} A_{ij} e^{-(x/v)(1/\tau_i)} \tag{8-13}$$

where n_i is the initial-state occupation number, A_{ij} is the transition probability from level i to level j, x is the distance downstream from the foil traveled by the particles, v their velocity, and τ_i the life time of the initial excited state. Equation 8-13 implies

$$\ln I_k = \ln I_{i0} - \frac{x}{v} \frac{1}{\tau_i} \tag{8-14}$$

from which the life time τ_i can be obtained by photoelectrically measuring the intensity of the emitted radiation as a function of distance from the foil. The beam-foil technique can give accurate results for almost all kinds of transitions in neutral and multiply ionized atoms. The reason is that the complications present in other methods do not appear here. Density estimates are of course unnecessary, whereas production of ionized atomic species is now easily accomplished.

It may be noted in this context that production of very highly charged positive ions and spectral measurements on them is now possible for the first time in the terrestrial laboratory. These measurements will facilitate the study of relativistic effects in atoms and also will enable us to establish accurate ionization potentials for all stages of ionization needed in calculating "experimental" correlation energies and binding energies of molecules.

Uncertainties arising from collisional deexcitations or recombinations are nonexistent in the beam-foil technique since the process occurs in high vacuum ($\sim 10^{-6}$ Torr) and absorption of any emitted light is negligible. Electric or magnetic fields can be maintained in small values. Thus the emitted light is caused almost entirely by free decay of excited states (Bickel 1968).

A limitation to this method for calculating absolute oscillator strengths is the existence of cascade effects for certain excited states. Although the life times are still measurable, branching ratios for the occurring transitions must be known from another source to obtain absolute transition probabilities. However for the transitions of the type

$$1s^2 2s^2 2p^m \rightarrow 1s^2 2s 2p^{m+1}$$

cascade processes are thought not to cause undue complications.

It is seen therefore that recent experimental methods can yield rather accurate multiplet oscillator strengths. Emission experiments seem to have uncertainties of about 15% to 20%, the phase shift measurements of about 10% to 15%, and the beam-foil technique produces results with a 5% to 15% error.

How does theory compare with these results? Can it yield results with the same accuracy? Traditional approaches are known not to succeed in this, and a new approach has been needed for some time.

Theoretical Methods

Different simplified approximations have been used in the past depending on the type of transitions considered. For example, if one looks at electric dipole transitions which are "one-electron jumps" between outer orbitals and neglects changes which are in the inner orbitals, one can approximate the matrix elements by one-electron integrals between the final and initial orbitals. Such an approximation is most applicable in the cases of alkali-like atoms and ions where there occur transitions of an outer electron from one subshell to another (e.g., $3s$–$3p$ in Na). These are cases where the "jumping" electron has a principal quantum number, in both the initial and final states, larger than those of the core electrons. However, core polarization can affect the oscillator strengths from 1% for Li to as much as 16% for the heavier alkali metals such as Cs (Hammeed, Herzenberg, James 1968). The wavefunctions used are based on the independent-particle model and can yield fairly good results. Bates and Damgaard (1949) have used what is known as the Coulomb approximation to obtain the necessary initial and final orbitals for a great number of transitions. In this approximation, asymptotic solutions of the Schroedinger equation are used to describe the outer electron moving in a potential caused by a screened nuclear charge. In spite of the oversimplicity of this model, the approximation has given good results for many cases because of the simpler nature of transitions of this kind and because of the semiempirical aspect of the method.

A more sophisticated independent-particle model is the Hartree–Fock method where the individual orbitals are obtained from the well-known coupled integro-differential Hartree–Fock equations. Calculations on the lithium isoelectronic sequence have shown that when the $(n-1)$ passive electrons form a closed-shell, self-consistent field, calculations of the initial and final orbitals of the jumping electron lead to quite good results for the oscillator strengths (Weiss 1963). However, as remarked above, with heavier alkalis, "core polarization" often makes the Hartree–Fock approximation inadequate even for such simple transitions.

For transitions involving equivalent electrons in open shells in either or both terms, e.g., the transitions in the ultraviolet spectrum of the type

$$1s^2 2s^2 2p^n \rightarrow 1s^2 2s 2p^{n+1}$$

correlation effects become exceedingly important, and the independent-particle models whether Hartree–Fock or the Coulomb approximation became entirely inadequate (Layzer and Garstang 1968). The Coulomb approximation is actually not applicable, while oscillator strengths obtained with Hartree–Fock wavefunctions are usually in disagreement with experiments in many instances by a factor of 2 or 3.

Another and better approximation that has been used is the Z-expansion method which applies to transitions in which the principal quantum number remains the same (Layzer and Garstang 1968). In this method, the energy and the wavefunctions are given by an expansion in term of powers of Z, the nuclear charge, with coefficients determined by perturbation theory. To the first order, a calculation allows mixing of nearly degenerate configurations and predicts the two leading terms in the energy expansion:

$$E = E_0 Z^2 + E_1 Z + E_2 + 0(Z^{-1}) + \cdots \tag{8-15}$$

and the leading term in the Z expansion of any other quantity (Layzer and Garstang 1968). Froese (1965) has carried out such calculations on first-row atoms. In a slightly modified approach, where a screening parameter is used in the Z expansion, Dalgarno and co-workers have considered the mixing of a few terms (Cohen and Dalgarno 1964; Crossley and Dalgarno 1965). The results of the above calculations, although improving on the Hartree–Fock (H-F) approximation, are generally still not in agreement with experiment. It is apparent that additional configurations and effects, supplementing those degenerate with the H-F wavefunction, must also be included. Weiss (1967) has considered more extensive wavefunctions using "superposition of configurations," which is the expansion of a wavefunction in infinitely many Slater determinants, for some transitions in C I and C II. A large number of configurations is considered—up to forty—expressed in terms of Slater determinants.

This method takes into account a larger portion of the correlation energy and, in the case of oscillator strengths, better agreement with experiment is achieved than the previously mentioned methods. The method of configuration interaction being a series expansion is, of course, in principle, capable of giving the correct results. However there is the crucial problem of which configurations to "mix" in order to obtain a good wavefunction suitable for calculations of oscillator strengths. All configurations important for determining the energy are not of equal importance in computing transition probabilities, which would seem to depend strongly only upon those configurations needed to specify an accurate charge distribution. Especially because superposition of configurations is in general slowly convergent, there is the likelihood of including many configurations which make little difference and of missing some which may be important. It is not a physical theory with a guiding principle as to significant effects, but a computational procedure.

Beyond the Hartree–Fock method, the exact wavefunction and other properties depend on electron correlation. The many-electron theory of Sinanoğlu and co-workers treats electron correlation systematically (Silverstone and Sinanoğlu 1966; Öksüz and Sinanoğlu 1969). The various kinds of correlation effects are now explicitly found and taken into account. They are physically and mathematically indicated by the theory. In the following discussion the main features of the Sinanoğlu MET theory are briefly reviewed and the oscillator strengths computed using it, for transitions of the type

$$1s^2 2s^2 2p^n \rightarrow 1s^2 2s 2p^{n+1}$$

are given and compared with other theoretical and experimental values.

Non-closed Shell Many-Electron Theory (NCMET)
Atomic Structure and Transitions

Sinanoğlu and co-workers in NCMET have shown that the exact nonrelativistic N-electron wavefunction can be expressed as

$$\psi = \phi_{RHF} + \chi_{int} + \chi_F + \chi_U \tag{8-16}$$

with all parts orthogonal to each other and $\langle \phi_{RHF} \phi_{RHF} \rangle = 1$. Here, ϕ_{RHF} is the restricted Hartree–Fock wavefunction, which in general has the form

$$\phi_{RHF} = \sum_{K \geqslant 1}^{\alpha} C_K \Delta_K \tag{8-17}$$

where Δ_K is a Slater determinant, and all the determinants in the sum belong to the same configuration of orbitals. As it is shown in the chapter by Sinanoğlu in the first volume of this book, the χ_{int} and χ_F are the nondynamical correlation effects which affect charge distributions the most, but they can be calculated completely in terms of only a finite configuration interaction (CI) expansion in terms of Slater determinants which involve, say, in the $1s^n2s^m2p^k$ states only the $1s$, $2s$, $2p$ Hartree–Fock orbitals and at most four new radial functions. The term χ_U, consists mainly of shorter range dynamical pair correlations and is the only effect that remains in closed-shell systems. The calculation of these would be slowly convergent and require an infinite series CI. All three kinds of correlation, the internal (int), semiinternal and polarization (F), and the all-external (U), are important in calculating energies of atomic systems and related quantities. However, in calculating transition probabilities, only the internal (int) and the semiinternal and polarization (F) correlation effects are taken into account. This is because, in analogy with the closed-shell systems where all-external correlations do not affect the Hartree–Fock charge distribution, the RHF wavefunction along with the nondynamical correlations can most accurately describe the charge distribution and therefore can be used to evaluate to high accuracy matrix elements of the transition operators. Therefore, in predicting the wavefunctions capable of yielding transition probabilities, the hypothesis is made that the all-external correlations are not important; a hypothesis borne out by the results. The wavefunctions to be used, then, are of the form

$$\psi' = \phi_{RHF} + \chi_{int} + \chi_F \tag{8-18}$$

As we have seen, such a type of wavefunction can be calculated by a finite and theoretically determined configuration-interaction calculation with no ambiguity as to which configurations to choose. Such calculations were first carried out for 113 states of the $1s^22s^n2p^m$ type of B, C, N, O, F, Ne, Na, and their ions by I. Öksüz and O. Sinanoğlu (1969). These wavefunctions contain all the nondynamical correlation effects among L-shell electrons. They are given as a finite linear combination of Slater determinants Δ_K constructed from n single-particle functions selected from an orthonormal set of Q orbitals $\varphi_1 \cdots \varphi_M \cdots \varphi_Q$. The first M spin orbitals define the H-F sea (for the first row $M = 10$). The remaining $(Q - M)$ orbitals, with at most four radial functions, are those that give the semiinternal f_{ij}; l and polarization functions, f_p. (See the chapter by Sinanoğlu in Volume I.) These new functions have only the s, p, d, f symmetries with radial functions closely approximated by $3s$-like, $3p$-like, $3d$-like and $4f$-like functions with optimized exponents. For all states γLSM_LM_s all the single-particle functions involved therefore have the form

$$\varphi_i = R_{l_i}(r)Y_{l_i}^{m_i}(\theta, \varphi)X_{m_{s_i}}(\sigma_z) \tag{8-19}$$

where Y and X are the normalized spherical harmonics and spin functions. The term $R_{l_i}(r)$ can be expanded as a sum of Slater-type orbitals (STO) but even one STO with an optimized exponent is quite sufficient. Using these wavefunctions, calculations for 29 transitions of the type

$$1s^2 2s^2 2p^n \rightarrow 1s^2 2s 2p^{n+1}$$

have been performed by Westhaus and Sinanoğlu (1969a). An important point to note in this work is that the "nonorthogonality" of orbitals belonging to different states is explicitly taken into account in evaluating the N-body integrals

$$\langle\psi_{\gamma LSM_LM_S}|\mathbf{O}|\psi_{\gamma'L'S'M_L'M_S'}\rangle = \sum_K \sum_{K'} C_K C_K' \int dx_1 \cdots dx_N \Delta_K^* \mathbf{O}_i \Delta_{K'}' \tag{8-20}$$

Thus the customary "frozen-core" approximation is abandoned in favor of a rigorous evaluation of these N-electron integrals.

In Table 8-1, the absorption oscillator strengths for the 29 ultraviolet transitions in the species C II, N I, N II, N III, O II, O IV, F II, Ne II, and Na III are presented. In column two, the oscillator strengths found in the National Bureau of Standards (NBS) Tables of May 1966 are shown. The next six columns contain results obtained with RHF and the MET wavefunctions of the Sinanoğlu theory. The symbols R, ∇ and $(R\nabla)^{\frac{1}{2}}$ denote the dipole-length, dipole-velocity, and the geometric mean (see Equation 8-12) operators respectively. The final column lists the available experimental data and their uncertainty. It is seen that, on the basis of comparison with experimental results, the MET values are superior to those tabulated in the NBS tables, although the latter often represent a substantial improvement upon the RHF results by including limited CI, which aims at taking into account part of what in MET is referred to as internal correlation. Clearly such limited conventional CI is not always sufficient to bring the calculated oscillator strengths into agreement with experiments—the remaining nondynamical correlations are also important. To illustrate this, we present in Table 8-2 the results of Cohen and Dalgarno (1964), who introduce internal correlation into the lower states by mixing configurations degenerate with the RHF wavefunction and compare them with the RHF, NBS, and MET values. A result obtained by A. Weiss (1967) using the "superposition of configurations" method is also given. It is clear from Table 8-2 that all nondynamical correlation effects including the semiinternal, as shown by the Sinanoğlu MET, need to be included in the wavefunctions to obtain accurate oscillator strengths. The agreement between the experimental data and the corresponding MET values is indeed an encouraging confirmation of the theory, particularly in the light of the substantial role that correlation effects play in determining these oscillator strengths. Thus, whenever experimental corroboration does not exist, MET-predicted values may be used with considerable confidence in atomic and astrophysical applications.

FORBIDDEN TRANSITIONS

The subject of forbidden transitions, important in atmospheric physics and astrophysical research, has been reviewed by R. H. Garstang in this volume and in references given therein. We shall discuss and give only the new results obtained by using the electron correlation theory and the many-electron wavefunctions described above (Nicolaides, Westhaus, and Sinanoğlu).

Electric quadrupole transition probabilities for the oxygen atom auroral green line $^1S_0-^1D_2$ ($\lambda = 5577$ Å) and the $^2P-^2D$, $^1S-^1D$ lines of oxygen ions, nitrogen, and nitrogen ions have been calculated with the above theory to obtain values that are more accurate than the existing values and also to establish, for the first time in a systematic way, the effect of electron correlation on forbidden electric quadrupole lines (Nicolaides, Westhaus, and Sinanoğlu).

The experimental results are not so accurate as the ones obtained for allowed dipole transitions. They are limited to the important O I $^1S_0-^1D_2$ auroral line. By studying several types of rapidly changing aurorae in which competitive factors, like collisional deactivation, vary, Omholt (1956 and 1959) has calculated an average transition probability of A (5577) $= 1.43 \pm 14\%$ sec^{-1}. Le Blanc, Oldenberg, and Carleton (1966) have studied the radiative decay of $^1S_0$ in the laboratory. They obtained a value at $A = 1.36 \pm 0.20$ sec^{-1}. McConkey and Kernahan (1969) have made an absolute intensity measurement. They give a value of $A = 1.0$ sec^{-1} with, however, several sources of error, making the result reliable only to within a factor of 2. A new experimental method is currently being applied by Dr. Corney and collaborators at Oxford.

Previous theoretical results were obtained using an independent particle model in which a single one-electron integral was calculated (Garstang 1951 and 1956). Our many-body calculations are the first of this kind. The transition probability for electric quadrupole radiation is

Table 8-1. Multiplet absorption strengths for 29 ultraviolet $1s^2 2s^2 2p^n \to 1s^2 2s 2p^{n+1}$ transitions obtained with the dipole-length velocity operators using RHF and MET wavefunctions. The MET results are from Westhaus and Sinanoğlu (1969).

	Transition	f^{NBS}	f_R^{RHF}	f_V^{RHF}	$f_{(RV)\ddagger}^{\text{RHF}}$	f_R^{MET}	f_V^{MET}	$f_{(RV)\ddagger}^{\text{MET}}$	$f^{\text{Experiment}}$
C II	$2p\,{}^2P \to 2p^2\,{}^2D$	0.27	0.263	0.262	0.262	0.125	0.134	0.129	0.114^a (10%)
N III	$2p\,{}^2P \to 2p^2\,{}^2D$	0.18	0.213	0.214	0.213	0.114	0.125	0.119	0.103^b (10%)
C IV	$2p\,{}^2P \to 2p^2\,{}^2D$	0.15	0.179	0.181	0.180	0.106	0.111	0.108	0.091^c (3%)
C II	$2p\,{}^2P \to 2p^2\,{}^2S$	0.059	0.070	0.042	0.054	0.122	0.121	0.121	
N III	$2p\,{}^2P \to 2p^2\,{}^2S$	0.11	0.056	0.035	0.044	0.085	0.084	0.084	
O IV	$2p\,{}^2P \to 2p^2\,{}^2S$	0.10	0.047	0.030	0.038	0.069	0.071	0.070	
C II	$2p\,{}^2P \to 2p^2\,{}^2P$	0.52	0.736	0.282	0.456	0.501	0.471	0.486	
N III	$2p\,{}^2P \to 2p^2\,{}^2P$	0.45	0.577	0.227	0.362	0.399	0.390	0.394	0.416^b (18%)
O IV	$2p\,{}^2P \to 2p^2\,{}^2P$	0.38	0.473	0.189	0.299	0.334	0.329	0.331	
N II	$2p^2\,{}^3P \to 2p^3\,{}^3D$	0.17	0.236	0.268	0.251	0.100	0.105	0.102	0.109^a (11%)
O III	$2p^2\,{}^3P \to 2p^3\,{}^3D$	0.15	0.200	0.225	0.212	0.100	0.104	0.102	0.101^b (6%)
N II	$2p^2\,{}^3P \to 2p^3\,{}^3P$	0.22	0.170	0.138	0.153	0.137	0.155	0.146	0.102^c (3%)
O III	$2p^2\,{}^3P \to 2p^3\,{}^3P$	0.18	0.143	0.117	0.129	0.127	0.135	0.131	0.131^b (6%)
N II	$2p^2\,{}^3P \to 2p^3\,{}^3S$	0.23	0.334	0.110	0.192	0.218	0.203	0.210	0.189^b (9%)
O III	$2p^2\,{}^3P \to 2p^3\,{}^3S$	0.19	0.272	0.092	0.158	0.183	0.173	0.178	
N II	$2p^2\,{}^1D \to 2p^3\,{}^1D$	0.45	0.651	0.310	0.449	0.314	0.327	0.320	

Ion	Transition								
O III	$2p^2 \, ^1D \rightarrow 2p^3 \, ^1D$	0.37	0.534	0.263	0.375	0.297	0.303	0.300	
N II	$2p^2 \, ^1D \rightarrow 2p^3 \, ^1P$	0.30	0.245	0.094	0.152	0.298	0.261	0.279	
O III	$2p^2 \, ^1D \rightarrow 2p^3 \, ^1D$	0.25	0.202	0.080	0.127	0.219	0.193	0.206	
N II	$2p^2 \, ^1S \rightarrow 2p^3 \, ^1P$	0.40	0.817	0.457	0.611	0.259	0.309	0.283	
O III	$2p^2 \, ^1S \rightarrow 2p^3 \, ^1P$	0.35	0.669	0.388	0.509	0.294	0.337	0.315	0.080[a] (10%)
N I	$2p^3 \, ^4S \rightarrow 2p^4 \, ^4P$	0.13	0.503	0.542	0.522	0.145	0.176	0.160	0.13[d]
O II	$2p^3 \, ^4S \rightarrow 2p^4 \, ^4P$	0.43	0.428	0.457	0.442	0.206	0.225	0.215	0.182[c] (3%)
O II	$2p^3 \, ^2D \rightarrow 2p^4 \, ^2D$	0.25	0.263	0.189	0.223	0.141	0.167	0.153	
O II	$2p^3 \, ^2P \rightarrow 2p^4 \, ^2S$	0.15	0.125	0.081	0.101	0.097	0.102	0.099	
O II	$2p^3 \, ^2P \rightarrow 2p^4 \, ^2D$	0.07	0.126	0.122	0.124	0.030	0.043	0.036	
F II	$2p^4 \, ^3P \rightarrow 2p^5 \, ^3P$	0.56	0.322	0.263	0.291	0.140	0.172	0.155	
Ne II	$2p^5 \, ^2P \rightarrow 2p^6 \, ^2S$	0.33	0.176	0.117	0.143	0.073	0.091	0.082	
Na III	$2p^5 \, ^2P \rightarrow 2p^6 \, ^2S$		0.155	0.105	0.128	0.077	0.090	0.083	0.035[e] > 0.055[f]

[a] Lawrence and Savage (1966) phase shift.
[b] Heroux (1967) beam-foil.
[c] Bickel (1967) beam-foil.
[d] Labuhn (1965) emission.
[e] Hinnov (1966) emission.
[f] Lawrence and Hesser, unpublished. See Hinnov (1966).

given by Shortley (1940) as

$$A_Q(\gamma LSJ \to \gamma'L'S'J') = \frac{1}{2J+1} \frac{1679.2 \times 10^{15}}{\lambda^5} S_Q(\gamma LSJ \to \gamma'L'S'J') \quad (8\text{-}21)$$

where A_Q is given in $\sec^{-1}$, λ in angstrom units, and S_Q is the "line strength" defined by

$$S_Q(\gamma LSJ \to \gamma'L'S'J') = \sum_{M_J M'_J} |\langle \gamma LSJM_J |Q| \gamma'L'S'J'M'_J \rangle|^2 \quad (8\text{-}22)$$

and given in atomic units. The electric quadrupole operator Q is given by:

$$Q = -\sum_i (\mathbf{r}_i \mathbf{r}_i - \tfrac{1}{3}\mathbf{r}_i^2 \mathscr{F}) \qquad (\mathscr{F} = \hat{\imath}\hat{\imath} + \hat{\jmath}\hat{\jmath} + \hat{k}\hat{k}) \quad (8\text{-}23)$$

the summation running over the electrons. The wavefunctions of the initial and final states are expressed in L-S coupling as a linear combination of Slater determinants. The N-body matrix elements can then be reduced ultimately to sums of products of N one-body matrix elements. As in the case of allowed transitions, the problem of nonorthogonality of the two sets of spin orbitals is explicitly taken into account. After we have obtained an answer for the matrix element evaluation in the LSM_LM_S scheme, a reduced matrix element, $\langle \gamma SL \| Q \| \gamma' S'L' \rangle$, can be deduced from it. Then, the desired S_Q in $SLJM$ scheme can be obtained, using the transformations given by Shortley (1940), which result into

$$S_Q(\gamma LSJ \to \gamma'L'S'J') = f_q(LSJ, L'S'J') |\langle \gamma SL \| Q \| \gamma' S'L' \rangle|^2 \quad (8\text{-}24)$$

where f_q is $\frac{1}{16}$ the coefficient of G^2, H^2, or I^2 in the table on page 253 of Condon and Shortley's *Theory of Atomic Spectra*. In obtaining the transition probability from the line strength S_Q, the experimental wavelengths have been used.

The NCMET results are given in Table 8-3 and are compared with the NBS results, which are based on theoretical calculations using Hartree–Fock wavefunctions. Results obtained by us while using just the restricted Hartree–Fock wavefunctions are also given. It is seen that inclusion of electron correlation ($\chi_{RHF} + \chi_{int} + \chi_F$) results in values smaller than the ones obtained by the use of just the H-F wavefunctions by 13% to 17%. This effect of electron correlation on the electric quadrupole transitions is less drastic than the effect on the allowed ones, as it has been suspected by Garstang. It is now possible to establish with certainty that the effect of electron correlation reduces the value obtained by H-F wavefunctions by about 15%. In any case, the results obtained with the new atomic structure theory with electron correlation may be taken as quite reliable

Table 8-2. The absorption-oscillator strengths (f) for the multiplets are computed using the dipole-length operator. The values listed in the f^Z column are obtained with restricted Hartree–Fock (RHF) wavefunctions, while those in the f^{NBS} column come from various calculations as indicated by the references. The values in the f^{MET} column are computed with wavefunctions of the present theory (MET).[h] The last column lists the available experimental results.

Species	Transitions	$\lambda(Å)$	f^{RHF}	f^Z	f^{NBS}	f^{MET}	$f^{Experiment}$
C II	$1s^22s^22p\ ^2P \rightarrow 1s^22s2p^2\ ^2D$	1335	0.263	0.204[g]	0.17[a1] (0.121)[a2]	0.125	0.114 (±.011)[b]
N II	$1s^22s^22p^2\ ^3P \rightarrow 1s^22s2p^3\ ^3D$	1085	0.236	0.192[g]	0.17[c]	0.100	0.109 (±.011)[b] 0.101 (±.006)[d]
N II	$1s^22s^22p^2\ ^3P \rightarrow 1s^22s2p^3\ ^3P$	916	0.170	0.213[g]	0.22[c]	0.137	0.131 (±.007)[d]
N II	$1s^22s^22p^2\ ^3P \rightarrow 1s^22s2p^3\ ^3S$	645	0.334	0.244[g]	0.23[c]	0.218	0.189 (±.016)[d]
N III	$1s^22s^22p\ ^2P \rightarrow 1s^2\ 2s2p^2\ ^2D$	991	0.213	0.167[g]	0.18[e]	0.114	0.103 (±.010)[d]
N III	$1s^22s^22p\ ^2P \rightarrow 1s^22s2p^2\ ^2P$	686	0.577	0.415[g]	0.45[e]	0.399	0.416 (±.075)[d]
O III	$1s^22s^22p^2\ ^3P \rightarrow 1s^22s2p^3\ ^3D$	834	0.200	0.162[g]	0.15[c]	0.100	0.102 (±.002)[f]
O III	$1s^22s^22p^2\ ^1D \rightarrow 1s^22s2p^3\ ^1D$	600	0.534		0.37[c]	0.297	—
O IV	$1s^22s^22p\ ^2P \rightarrow 1s^22s2p^2\ ^2D$	789	0.179	0.141[g]	0.15[e]	0.106	0.091 (±.002)[f]

[a1] Weiss (in NBS Tables). [a2] Weiss (1967).
[b] Lawrence and Savage (1966).
[c] Bolotin, Levinson, and Levin (1956).
[d] Heroux (1967).
[e] Bolotin and Yutsis (1953).
[f] Bickel (1967).
[g] Cohen and Dalgarno (1964).
[h] Westhaus and Sinanoğlu (1969b).

Table 8-3. Calculated electric quadrupole transitions in the O I, O II, O III, N I, N II ions. The NCMET wavefunctions used, include all the nondynamical correlation effects. The NBS values are also given for comparison. The χ_{non-D} is χ_{int+F} (cf. text).

	Transitions	NCMET				NBS	
		A_Q		S_Q		A_Q	S_Q
		RHF	RHF + χ_{non-D}	RHF	RHF + χ_{non-D}		
O I	$^1D_2-^1S_0$	1.421	1.183	4.567	3.801	1.34[a]	4.31
O II	$^2D_{5/2}-^2P_{3/2}$	0.1060	0.0915	5.3039	4.5809	0.106	5.30
	$^2D_{5/2}-^2P_{3/2}$	0.0606	0.0523	1.5154	1.3088	0.0610	1.52
	$^2D_{3/2}-^2P_{1/2}$	0.0449	0.0388	2.2731	1.9632	0.0450	2.29
	$^2D_{3/2}-^2P_{3/2}$	0.0902	0.0779	2.2731	1.9632	0.0900	2.27
O III	$^1D_2-^1S_0$	1.824	1.654	1.717	1.557	1.60	1.51
N I	$^2D_{5/2}-^2P_{3/2}$	0.0590	0.0489	17.0819	14.1545	0.054	15.6
	$^2D_{5/2}-^2P_{3/2}$	0.0337	0.0279	4.8805	4.0441	0.0308	4.45
	$^2D_{3/2}-^2P_{1/2}$	0.0251	0.0208	7.3208	6.0662	0.0230	6.7
	$^2D_{3/2}-^2P_{3/2}$	0.0504	0.0417	7.3208	6.0662	0.0460	6.7
N II	$^1D_2-^1S_0$	1.240	1.082	4.661	4.065	1.08	4.06

[a] This value is an average of the theoretical value by Garstang (1951 and 1956) and the experimental one by Omholt (1956 and 1959).

for use in various applications as in connection with the nebulas, the study of aurorae, and the airglow.

The work that led to this chapter was supported by a grant to O.S. by the U.S. National Science Foundation.

REFERENCES

Bashkin, S. 1968. *Beam-Foil Spectroscopy*. New York: Gordon and Breach.
Bashkin, S., et al. 1964. *Phys. Letters* 10:63.
Bates, D. R., and Damgaard, A. 1949. *Phil. Trans. Roy. Soc.* London A242:101.
Bickel, W. S.
    1967. *Phys. Rev.* 162:7.
    1968. *Appl. Opt.* 7:2367.
Bolotin, A. B.; Levinson, I. B.; and Levin, L. I. 1956. *Sov. Phys.—JETP* 2:391.
Bolotin, A. B., and Yutsis, A. P. 1953. *Sov. Phys.—JETP* 24:537.
Chandrasekhar, S. 1945. *Astrophys. J.* 102:223.
Cohen, M., and Dalgarno, A. 1964. *Proc. Roy. Soc.* London A280:258.
Crossley, R., and Dalgarno, A. 1965. *Proc. Roy. Soc.* London A286:510.
Foster, E. W. 1964. *Repts. Progr. Phys.* 27:469.
Froese, C. 1965. *Astrophys. J.* 141:1206.
Garstang, R. H.
    1951. *Monthly Notices Roy. Astron. Soc.* 111:115.
    1956. In *The Airglow and Aurorae*, ed. F. B. Armstrong and A. Dalgarno. London, U.K.: Pergamon.
Hammeed, S.; Herzenberg, A.; and James, N. C. 1968. *J. Phys. B.* 1:822.
Heroux, L. 1967. *Phys. Rev.* 153:156.
Hinnov, E. 1966. *J. Opt. Soc. Am.* 56:1179.
Labuhn, F. 1965. *Z. Naturforsch* 20a:995.
Lawrence, A. M., and Savage, B. D. 1966. *Phys. Rev.* 141:67.
Layzer, D., and Garstang, R. H. 1968. *Ann. Rev. Astron. Astrophys.* 6:449.
LeBlanc, F. J.; Oldenberg, O.; and Carleton, N. P. 1966. *J. Chem. Phys.* 45:2200.
McConkey, J. W. and Kernahan, J. A. 1969. *Planet Space Sci.* 17:1297.
Nicolaides, C.; Westhaus, P.; and Sinanoğlu, O. *Phys. Rev.* A, 4:1400 (1971)
Öksüz, I., and Sinanoğlu, O. 1969. *Phys. Rev.* 181:42, 54.
Omholt, A. 1956. In *The Airglow and the Aurorae*, ed. F. B. Armstrong and A. Dalgarno. London, U.K.; Pergamon.
Omholt, A. 1959. *Geofys. Publi. Norske Videnskaps-Akad. Oslo* 21:1.
Shortley, G. H. 1940. *Phys. Rev.* 57:225.
Silverstone, H. J., and Sinanoğlu, O. 1966. *J. Chem. Phys.* 44:1899, 3608.
Westhaus, P., and Sinanoğlu, O.
    1969a. *Phys. Rev.* 183:56.
    1969b. *Astrophys. J.* 157:997.
Weiss, A. W.
    1963. *Astrophys. J.* 135:1262.
    1967. *Phys. Rev.* 162:71.
Wiese, W. L. 1968. *Methods Exp. Phys.* 7a:117.

Subject Index